PARENTERAL NUTRITION IN INFANCY AND CHILDHOOD

ADVANCES IN EXPERIMENTAL MEDICINE AND BIOLOGY

Editorial Board:

Nathan Back	State University of New York at Buffalo
N. R. Di Luzio	Tulane University School of Medicine
Bernard Halpern	Collège de France and Institute of Immuno-Biology
Ephraim Katchalski	The Weizmann Institute of Science
David Kritchevsky	Wistar Institute
Abel Lajtha	New York State Research Institute for Neurochemistry and Drug Addiction
Rodolfo Paoletti	University of Milan

Recent Volumes in this Series

Volume 38
HUMAN HYPERLIPOPROTEINEMIAS: Principles and Methods
Edited by R. Fumagalli, G. Ricci, and S. Gorini • 1973

Volume 39
CURRENT TOPICS IN CORONARY RESEARCH
Edited by Colin M. Bloor and Ray A. Olsson • 1973

Volume 40
METAL IONS IN BIOLOGICAL SYSTEMS: Studies of Some Biochemical and
Environmental Problems
Edited by Sanat K. Dahr • 1973

Volume 41A
PURINE METABOLISM IN MAN: Enzymes and Metabolic Pathways
Edited by O. Sperling, A. De Vries, and J. B. Wyngaarden • 1974

Volume 41B
PURINE METABOLISM IN MAN: Biochemistry and Pharmacology of Uric Acid Metabolism
Edited by O. Sperling, A. De Vries, and J. B. Wyngaarden • 1974

Volume 42
IMMOBILIZED BIOCHEMICALS AND AFFINITY CHROMATOGRAPHY
Edited by R. B. Dunlap • 1974

Volume 43
ARTERIAL MESENCHYME AND ARTERIOSCLEROSIS
Edited by William D. Wagner and Thomas B. Clarkson • 1974

Volume 44
CONTROL OF GENE EXPRESSION
Edited by Alexander Kohn and Adam Shatkay • 1974

Volume 45
THE IMMUNOGLOBULIN A SYSTEM
Edited by Jiri Mestecky and Alexander R. Lawton • 1974

Volume 46
PARENTERAL NUTRITION IN INFANCY AND CHILDHOOD
Edited by Hans Henning Bode and Joseph B. Warshaw • 1974

Volume 47
CONTROLLED RELEASE OF BIOLOGICALLY ACTIVE AGENTS
Edited by A. C. Tanquary and R. E. Lacey • 1974

PARENTERAL NUTRITION IN INFANCY AND CHILDHOOD

Edited by

Hans Henning Bode
Department of Pediatrics, Harvard Medical School,
Massachusetts General Hospital, Shriners Burns Institute
Boston, Massachusetts

and

Joseph B. Warshaw
Departments of Pediatrics, Obstetrics, and Gynecology
Yale University School of Medicine
New Haven, Connecticut

PLENUM PRESS • NEW YORK AND LONDON

Library of Congress Cataloging in Publication Data

Main entry under title:

Parenteral nutrition in infancy and childhood.

(Advances in experimental medicine and biology, v. 46)

Based in part on the papers of the International Symposium on Parenteral Nutrition in Infancy and Childhood, held in Boston in 1972.

Includes bibliographies.

1. Parenteral therapy—Congresses. 2. Pediatrics—Congresses. I. Bode, Hans Henning, ed. II. Warshaw, Joseph B., 1936- ed. III. Series. [DNLM: 1. Parenteral feeding—In infancy and childhood—Congresses. W1AD559 v. 46 1972 / WB410 I64p 1972]
RJ53.F5P37   1972   615'.63   74-6060
ISBN 0-306-39046-9

Based in part on papers of the International Symposium on Parenteral Nutrition in Infancy and Childhood held at Boston, Massachusetts, August 29-31, 1972

© 1974 Plenum Press, New York
A Division of Plenum Publishing Corporation
227 West 17th Street, New York, N.Y. 10011

United Kingdom edition published by Plenum Press, London
A Division of Plenum Publishing Company, Ltd.
4a Lower John Street, London W1R 3PD, England

All rights reserved

No part of this book may be reproduced, stored in a retrieval system, or transmitted, in any form or by any means, electronic, mechanical, photocopying, microfilming, recording, or otherwise, without written permission from the Publisher

Printed in the United States of America

PREFACE

"A tranquil mind puts flesh on a man"
                                English proverb

After a period of relative neglect, nutrition as a medical science is now an area of great clinical and investigative activity. This renewed interest in clinical nutrition derives in large part from observations suggesting that early nutritional deprivation not only interferes with the maintenance of health, growth and resistance to disease but if present during critical periods of central nervous system development may also cause permanent impairment of intellectual capacity.

Studies on brain development during malnutrition have continued to demonstrate the vulnerability of the developing brain to nutritional insult. Winick (1968) has emphasized that nutritional deficiency occurring while cells of the central nervous system are actively dividing results in a permanent decrease in central nervous system cell number. Later nutritional deficiency which results in decrease in cell size appears to be recoverable. Perhaps even more important than effects of malnutrition on brain cell number is the effect on brain protein synthesis and myelination. As different regions of the brain grow at different rates and human cerebellar and cerebral cell number increase for the first few months of life, newborn nutritional deficiency may compromise brain development. Dobbing (1973) has focused attention on the vulnerability of the brain to nutritional insult during the brain growth spurt which occurs around the time of birth. In the human, this period extends throughout the third trimester of pregnancy and into the second postnatal year.

The full implication of nutritional deprivation on the development of intelligence is not entirely clear. However, the data available do suggest a permanent decrease in intellectual function in infancy (Cravioto, J. and DeLicardie, 1971; Cabak, V. and Majanvik, 1965; Winick, M. and Rosso, P., 1969). Thus, great demands are placed on the pediatricians and obstetricians who supervise the health of pregnant mothers and infants to eliminate subnormal substrate provision to the human fetus and the newborn during this critical period when the infant is at its greatest risk.

Worldwide, the most frequent cause of infant malnutrition is, unfortunately, still the limited availability of foodstuffs to both pregnant mothers and newborn infants. However, even in the more affluent societies malnutrition is frequently a consequence of utero-placental insufficiency, prematurity or neonatal starvation due to surgical procedure, congenital anomalies or other forms of distress which prevent adequate oral feeding. Provision of calories by means other than oral feeding are important not only during critical periods of early brain development but may also be necessary in the older infant or child with chronic disease associated with severe nutritional impairment or in children requiring good nutrition for optimal recovery from an acute ailment. Dudrick's work with intravenous feeding has paved the way for an aggressive nutritional approach to infants malnourished or at risk of becoming malnourished. Dudrick showed that beagle puppies receiving a hypertonic mixture of glucose, protein hydrolysate, minerals and vitamins by a central vein could maintain normal growth and development over prolonged periods of time. This technique of intravenous feeding was adapted to human use and allowed a premature newborn with intestinal atresia to sustain normal growth and development for a period of 44 days. Since this original case, total intravenous nutrition has been successfully administered to hundreds of infants, and the technique now occupies a prominent place in the therapy of the nutritionally endangered child. The technical details, the range of indications, and the modes of intravenous feeding are being refined, especially in the case of infants and children. New developments and recent experience have accumulated at a rapid pace. For this reason, we arranged a Symposium on Parenteral Nutrition in Children in the Fall of 1972. This volume is an outgrowth of that conference. The editors are grateful to J. Pfrimmer Company, Norwich Pharmacal Company and Vitrum Company for their support which made the conference and this volume possible.

Some of the studies reported on at the conference were controversial, and even though the editors and many of the participants did not agree with some of the views expressed, we felt that the data presented had created sufficient stimulation to justify publication. For example, the very aggressive approach for treatment of the fetus subjected to utero-placental insufficiency outlined by Dr. Heller opens a completely unexplored area of treatment and preventive medicine. Much more research in this area is needed and should probably be performed in primates. The varying incidences of catheter sepsis in different centers clearly emphasizes the importance of aseptic techniques and supervision and improvement of these techniques should probably take precedence over the prophylactic administration of amphotericin B as suggested by Dr. Brennen.

In this volume we have attempted to provide information concerned with biochemical and metabolic adaptations during development. We have organized this book so that the more basic chapters dealing with nutritional and metabolic development are presented early so as to provide a foundation for later chapters dealing with technique and application. It is our belief that only through an appreciation and understanding of biochemical and metabolic development can we rationally approach nutritional therapy of infants and children.

Hans H. Bode

Joseph B. Warshaw

REFERENCES

Cabak, V. and Majanvik, R. Arch. Dis. Childh., 40:532, 1965.

Carvioto, J. and DeLicardie, E. R. Nutr. Reviews, 29:107, 1971.

Dobbing, J. Lipids, Malnutrition and the Developing Brain. Ciba Foundation Symposium. Associated Scientific Publishers. p. 1, Amsterdam, 1973.

Winick, M. Pediat. Res., 2:352, 1968.

Winick, M. and Rosso, P. Pediat. Res., 3:181, 1969.

CONTENTS

	page
The history of parenteral alimentation by H. H. Bode	1
The regulation of energy intake by developing and adult animals by R. A. McCance	4
Factors influencing amino acid utilization by H. N. Munro	11
Amino Acid Requirements in childhood by D. M. Hegsted	27
Intravenous carbohydrate tolerance in infancy by G. W. Chance	38
Substrate supply and utilization in various conditions by J. F. Biebuyck	54
The utilization of xylitol, fructose and sorbitol by H. Förster	71
Fatty acid oxidation during development by J. B. Warshaw	88
Utilization and tolerance of intravenous fat emulsions by R. P. Geyer	98
Alcohol metabolism during development by E. Mezey	112
Fluid and electrolyte requirements and tolerance by J. D. Crawford	119
Trace elements and vitamins by H. L. Greene, M. Hambidge and Y. F. Herman	131

CONTENTS

	Page
The rapid rehabilitation of severely undernourished children by E. M. Widdowson	146
Technique of total parenteral nutrition in infants by S. J. Dudrick, B. V. MacFadyen and R. W. Winters	151
Postoperative parenteral feeding of neonates: Peripheral vein infusion technique, fat administration and metabolic studies by H. C. Børresen, R. Bjordal and O. Knutrud	165
Controlled parenteral nutrition of premature infants by P. Jurgens, D. Dolif, C. Panteliadis and C. Hofert	178
Total intravenous alimentation in low birth weight premature infants by W. C. Heird, J. M. Driscoll and R. W. Winters	199
Intrauterine amino acid feeding of the fetus by L. Heller	206
The role and effect of parenteral nutrition on the liver and its use in chronic inflammatory bowel disease in childhood by M. I. Cohen, S. J. Boley, F. Daum, I. F. Litt and S. K. Schonberg	214
Parenteral nutrition of renal disease by J. E. Fischer	225
Parenteral nutrition in critical illness by J. T. Herrin	231
PARENTERAL NUTRITION IN CHILDREN WITH BURNS	239
Experience in the Shriners Burns Institute in Cincinnati by M. P. Popp, E. J. Law and B. G. MacMillan	240
Experience in the Shriners Burns Institute in Boston by A. Antoon and H. H. Bode	247
Experience with central venous catheters by C. Burri and H. H. Pässler	250
Metabolic complications of total parenteral nutrition by W. C. Heird, R. W. Winters and S. J. Dudrick	256
Infection in association with intravenous feeding by M. F. Brennan	269

CONTENTS

	Page
Contributors	279
Author Index	283
Subject Index	291

THE HISTORY OF PARENTERAL ALIMENTATION

Hans H. Bode

Pediatrician and Endocrinologist, Children's Service,
Massachusetts General Hospital and Shriners Burns
Institute, Harvard Medical School, Boston, Mass.

Ever since William Harvey announced his theory of circulation in 1615, man has attempted to cure disease by cleansing or enriching the blood with infusion of intravenous remedies. Among the first to investigate this means of therapy was another famous Englishman, Sir Christopher Wren. He performed intravenous infusion in animals and discussed the benefits of blood transfusion, a technique that was undertaken in animals by Richard Lower in 1665 and ten years later by Jean Baptiste Denis who transfused sheep blood into humans. Fortunately, when these early studies met with little success or frequently even with disaster, the initial enthusiasm for human experimentation declined. Gauthier invented a technique for the distillation of water in 1717, and sixteen years later Stephen Hales produced "dropsy" in humans by infusing water intravenously. The study of metabolism was initiated by Lavoisier later in that century, and in 1792 glucose was discovered by Lobowitz.

The first experience with saline infusion was obtained during a cholera epidemic and reported by Latta in 1831. By 1850 hypodermic injection had been employed clinically by Rynd and Pravaz had introduced the hypodermic syringe. At about the same time Claude Bernard was investigating the effects of intravenously infused proteins and carbohydrate in animals. All of these studies were, however, destined to be only partially successful until the importance of microorganisms was recognized and microbiology was introduced into medicine by the work of Pasteur, Koch and others in the latter half of the last century. The astute clinical observations and the appreciation of antiseptic techniques by Semmelweis (1847) and Lister (1867) led to a marked reduction in morbidity and mortality of maternity and surgical patients. Landois's observation of hemolysis (1875) induced by infusion of alien blood which 38 years later led to Landsteiner's discovery of the blood groups, and Emil Fischer's ingenious studies of carbohydrates and amino acid analysis and synthesis were other important factors which opened the door for the rapid progress in intravenous therapy during the last hundred years.

Based on this newly acquired knowledge, Abderhalden et al (1909) designed the first synthetic diet which they successfully fed to a 9-year-old girl. After the First World War when the importance of pyrogen free infusion, need for vitamins, and the requirements and limits of electrolytes and water were elucidated, research in parenteral nutrition concentrated on the development of proper substances and techniques.

The recognition of protein wasting during starvation and especially during stress had led to the infusion of protein hydrolysates during the First World War. Shortly thereafter, Yamakawa began his investigations with intravenous fat emulsions. The beneficial effect of casein hydrolysate infusion on nitrogen balance in children was reported in 1939 (Shohl et al and Farr et al), and the studies on fat infusion were greatly advanced by the investigations of Geyer and his associates (1949). From then on progress was very rapid, but complete parenteral nutrition, although accomplished over long periods by few investigators (Coats, 1966), was limited by the excessive volume of water it required. This problem was partially solved 15 years ago in the United States with the commercial availability of fat emulsion. However, the initial cotton seed preparation caused several side effects that led to its withdrawal from the market in 1964. Fat emulsions manufactured from soybean oil in Europe proved safer and are still widely used outside the United States. The last barrier was overcome in 1968 with Dudrick's development of a technique that permitted central venous infusion of hypertonic solutions and delivery of nutrients sufficient to maintain normal growth and development even during infancy.

For the past decade parenteral alimentation has been advanced greatly; techniques and nutrient solution have been refined and the range of indications expanded. Intravenous nutrition has had a major impact on the prognosis of the chronically ill child and hopefully will prevent neurological impairment caused by nutritional deficits in the small newborn.

REFERENCES

1. Abderhalden, E., Frank, F. and Schittenlehn, A. Hoppe-Seylers, Z. Physiol. Chem. 63:215, 1909.

2. Coats, D.A. Anaesthes. Wiederbel. 6:142, 1966.

3. Dudrick, S.Y. In: Intravenous Hyperalimentation, G.S.M. Cowan and W.L. Scheetz, Eds. Philadelphia, Lea and Febiger, p. v, 1972.

4. Farr, L.E. and Mac Fadyen, D.H. Proc. Soc. Exp. Biol.(NY) 42:444, 1939.

5. Garrison, F.H. An Introduction to the History of Medicine. Philadelphia, W.B. Saunders Co., p. 809, 1929.

6. Geyer, R.P. Matthews, L.W. and Stare, F.J. J. Biol. Chem. 180:1037, 1949.

THE REGULATION OF ENERGY INTAKE BY DEVELOPING AND ADULT ANIMALS

R. A. McCance

Sidney Sussex College

Cambridge, England

I always advise my male acquaintances that if they want to brag about the size of their first born at birth, they should marry a large woman and feed her well during pregnancy, but if they want him to double his birth weight quickly, they should arrange to have him born prematurely.

This is not all such nonsense as it sounds. The rate at which all healthy organisms double their weight per unit of time falls from conception to maturity, and consequently, a small but healthy baby will double birth weight more quickly than the one that weighs 4.5 kg at birth and wins the prize at the parish baby shows.

THE FOETUS AND NEWBORN

The size of a baby at birth depends upon its genetic potential and its food supply. The father shares the responsibility for the former, but the nutrition of the foetus depends upon the supply of blood to the uterus and placenta, and this is where the importance of the large woman and her uterus comes in.

The evidence we have goes to show that the foetus takes all the nutrients and energy it can from the placental blood flow and grows accordingly. Failures to get enough are seen in such conditions as the runt pig, the offspring of undernourished ewes, multiple versus singleton products of conception in sheep and guinea pigs and the baby that is too small for its gestational age. The runt pig, for example, may be little more than a quarter the size of its well grown littermate at birth (Widdowson, 1971), yet reach a weight not much short of it by the age of three years, thus showing that the genetic potential was there. The same is true of the subsequent growth of twin lambs, and much of Hammond's (1961) work on the cross breeding of large and small strains of the same species probably showed the same thing (Walton and Hammond, 1938; McCance, 1962).

Animals born in an undeveloped state like the rat also appear to have very little control over their food intake for 10-14 days after birth and, by varying their plane of nutrition at this time, can be made not only to grow at very different rates but even to become fat (Knittle and Hirsch, 1968). Even animals born in a more highly developed state do not have the same curbs on their food intake as the adults do. "Rumen fill" and abdominal fat do not begin to limit the intake of food in newborn ruminants until after weaning. This is the age too at which dietary preferences for very high fat intakes can be established in the rat which produce immensely fat animals weighing 600-700 g by the time they are 6 months old (Mickelsen et al., 1955; Miller and Parsonage, 1972).

The human baby is sufficiently developed at birth to be able to vary, within limits, its food intake according to its needs: boys, for example, take more than girls from birth (Widdowson, 1971). Moreover, for at least a time, intake of an infant is limited by the bulk of the food presented to it and the capacity of its stomach and alimentary canal to hold it (Fomon, 1969).

It is also possible that at this age, and certainly a little later, many children will accept a greater energy intake than they require for their capacity to grow if it is suitably presented to them, and become fat in consequence (Widdowson, 1955).

ADULTS

When growth has ceased, the balance between the intake and expenditure of energy is generally very exact, and the general principles are well established. The expenditure is dictated by the internal requirements of the body, the activities of the individual and also by the environmental temperature, although the internal combustion engine, the clothiers and central heating have now succeeded in making us relatively independent of the last two. Man's intake of energy as food is normally well adjusted to his expenditure provided the food is available, but not from day to day, only over a week or more. This balance between the intake and expenditure of energy can be maintained in most of the animals investigated, including ourselves, in spite of considerable variation in the bulk and palatability of the diet, and this characteristic has been described, originally in rats, by saying that the animals "eat for calories." There are, however, always limits beyond which the relationship breaks down and, although these are interesting and instructive, we must consider the various theories which have been advanced to explain why there is a balance at all.

There have been three theories. They all depend upon the peripheral - central integration which takes place within the hypothalamus and brain stem. The first, which dates from 1948, was thermostatic, but this was criticised soon after it was put forward and replaced by a glucostatic theory. The basis of this was to suppose that there was a desire to eat when the sugar in the blood or sensitive cells was low, e.g., after being given insulin or alcohol, and a feeling of satisfaction when it rose. This is a good theory so far as it goes, and has been refined by recent work going to show that the real stimulus to eat does not originate in a fall in the level of circulating glucose so much as a failure to provide enough metabolisable glucose for certain sensitive cells in the lateral hypothalamus. The theory will explain the feeling of hunger before meals and of repletion afterwards, and a glycopeptide which produces anorexia, recently found in the serum of satiated rats, extends the idea of a water-soluble short term "stat" in an interesting way. This theory, however, can have no place in explaining the maintenance of a constant body weight for months or years.

The third and most satisfactory theory depends upon some compound or system which can monitor the amount of fat in the body and vary the intake of energy accordingly. A steroid with differential solubilities in water and fat has been suggested and this is a highly ingenious thought, but still, there are grave difficulties about it which its proposers themselves realise. The number of fat cells in the body, the amount to which the fat cells are distended or their overall surface area are all involved in possible variants of this theory in that they would monitor the amount of fat in the body and its fluctuations over long periods of time.

None of these theories, even if the details had all been worked out, explain all the facts. Thus (a) The growing animal can never be in energy balance. It must eat enough food every day, not only for the maintenance of the status quo, but to provide enough protein and energy for its growth until tomorrow. The glucostatic theory probably comes nearest to doing this. (b) They do not explain the increased intake of protein and energy - yielding materials during a physiological human pregnancy which considerably exceed the amounts required for the growth of the foetus, uterus, breasts, etc. and result in the deposition of fat and lean tissue in the general 'soma' of the mother. If this takes place under the influence of progesterone, as it does in rats, but not apparently in all species, this is a serious objection to any lipostatic theory based on a steroid. (c) If persons are transported to a hot environment, their expenditure and intake of energy tend to fall concurrently, but none of the theories explain why the balance is upset if the expenditure of energy is maintained at its original level. (d) The sensors of the body are not very alert about a rising percentage of fat in it and often fail to take the necessary action to reduce it, but they tend to become exceedingly active if there is any suggestion of undernutrition, and the healthy animal then does all it can to restore the percentage of fat in its body to its previous level even if this was an undesirably large one. Unbelievably large amounts of food may be consumed in such circumstances. Widdowson, for example, found that 6 undernourished elderly men 61 to 80 years old, over a period of 8 weeks, managed to consume enough food to provide themselves with 5617 kcal/day. A small and very undernourished girl in Uganda, 8½ months old and weighing only 2.5 kg, has been demonstrated to double this weight in the first seven weeks after admission and to consume over 250 kcal/kg day during the second and fifth week. The resistance of the person or animal who is overweight to having his weight reduced is all part and parcel of the same thing. This is all difficult to explain on the basis of any of the conventional "stat" mechanisms because thermostats and other physical regulators prevent a departure from the set range equally sharply in either direction, whereas the mammalian "appestat" is much more tolerant about too large an intake of energy than it is about one too small.

THE ROLE OF THE HYPOTHALAMUS

It is now well known that bilateral puncture of the ventromedial hypothalamic nucleus of an adult rat, and other animals, produces voracious eating for a time and a gross increase in the weight of the body, culminating in the establishment of a new plateau for it. A similar effect can be produced in mice by drugs. The hyperphagia is usually associated with reduced physical activity, and progesterone produces these effects in the females of some species. A reluctance to take exercise is rather characteristic of fat men and women, but to what extent the hypothalamus comes into this we do not know. It is largely by taking less exercise, however, that so many fat people have been able to persuade themselves and other people that they are virtually living on air. The hypothalamus, however, is much too complicated an organ for localised foci in it to be regarded as "lipostats." Puncture of the ventromedial hypothalamic nucleus may make an animal eat voraciously, but only for a time, and once it has reestablished control of its appetite and body weight, it is again prepared to "eat for calories," although not so fiercely as before.

Furthermore, puncture of a more lateral nucleus produces anorexia. This can be overcome by a palatable diet after a time, but the animal always remains choosy about its food.

This work on destructive lesions of the hypothalamus and brain stem has been extended by more refined investigations involving the localised introduction of transmitter substances and drugs by single injections or indwelling cannulae (Leibowitz, 1970).

VOLUNTARY AND BEHAVIORIAL ASPECTS

All living animals, whether they have a hypothalamus or not, possess what may be termed a "drive to eat," and in all the higher ones, the expenditure of energy is the initiating cause. The extent of the energy debt, moreover, determines the intensity of the drive and, furthermore, the amount consumed at each meal. This is one reason why the amount eaten each day begins to fall off as the original composition of the body is restored. Age, moreover, comes into this. The newborn rat will eat far more per kg per day in response to a limited energy debt than it will ever do later in life. A tiny marasmic child can eat 300 kcal per kg per day, but by 4 years of age, an equally wasted child is not likely to take more than about 150 kcal per kg per day, and the elderly men of 70 studied by Widdowson (1951) were unable to cope with more than 100 kcal per kg.

The drive to eat always tends to exceed the animals immediate needs for energy, particularly during growth, and it is only as growth slows down that the ventromedial hypothalamic nucleus of the rat really begins to exert its effect. The problems to be solved, therefore, are: (a) Why and how does the expenditure of energy actuate the drive to eat? We know something about the fringes of this problem; some tastes and smells increase the drive and so do movements of the empty stomach and intestine, low blood sugars, insulin, alcohol etc., but on what these act we seldom really know. (b) What are the mechanisms which stop an animal eating at each meal and over the day as a whole, even if it is still in considerable energy debt? Hard food and carious and loosening teeth may make it impossible for an animal or a man to satisfy his energy requirements, and strange, unpalatable food may do the same. If the diet is palatable, however, sensations known to us as satiety are the usual check. A single food may quickly produce a feeling of satiety at any one meal if this is all there is to eat, but this kind of satiety can be overcome in man by a succession of foods with different tastes and flavours, and elaborate menus are one of the reasons for obesity in prosperus communities. A succession of foods, however, sometimes has its therapeutic rewards, even in such unpromising animals as ailing pigs. Distention and discomfort must follow if the organism disregards the initial sensations of satiety and these are the feelings which no doubt ultimately prevent rats and other animals from meeting their demand for energy on food diluted with cellulose. They apply, moreover, in a very practical way to ruminants in whom "rumen fill", pregnancy and intra-abdominal fat may all limit the intake of food, even to less than they require to meet their expenditure of energy. These physical forces may be the ones that ultimately limit the food intake of the punctured rat, and its well known reluctance to face unpalatable food after it has re-restablished control of its weight may well be another facet of it.

It is a sad if a sobering thought that regulation of the intake and output of food in man differs so little from the same processes in other animals - but it does make its investigation easier!

REFERENCES

This article is based upon a paper to be published shortly by the same author in Nutrition Abstracts and Reviews. This will contain a full bibliography and the references cited below are merely points not mentioned in the forthcoming paper.

Foman, S.J., Filer, L.J., Thomas, L.N., Rogers, R.R. and Prolsch, A.M. Relationship between formula concentration and rate of growth of normal infants. J. Nutr., 98, 241-254, 1969.

Hammond, J. The effect of nutrition on the stage of development of the young at birth in farm animals. See Somatic Stability in the Newlyborn, (eds. Wolstenholme, G.E.W. and O'Connor, M.) pp. 5-15, Ciba Foundation Symposium. J. and A. Churchill, London, 1961.

Knittle, J.L. and Hirsch, J. Effect of early nutrition on the development of rat epididymal fat pads: Cellularity and metabolism. J. Clin. Invest., 47, 2091-2098, 1968.

Leibowitz, S.F. Hypothalamic beta-adrenergic "satiety" system antagonizes an alpha-adrenergic "hunger" system in the rat. Nature (London), 226, 963-964.

McCance, R.A. Food, growth and time. Lancet, 611-626 and 671-676, 1962.

Mickelsen, O., Takahashi, S. and Craig, C. Experimental obesity: 1. Production of obesity in rats by feeding high fat diets.

J. Nutr., 57, 541-554, 1955.

Miller, D.S. and Parsonage, S.R. The effect of calorie density on energy utilisation. Proc. Nutr. Soc., 31, 31A-32A, 1972.

Walton, A. and Hammond, J. The maternal effects on growth and conformation in Shire-horse Shetland-pony crosses. Proc. Roy. Soc. B., 125, 311-335, 1938.

Widdowson, E.M. The response to unlimited food. See Studies of undernutrition, Wuppertal 1946-9. Med. Res. Coun. Spec. Rep. Ser., 275, 313-345, 1951. H.M. Sta. Off., London.

Widdowson, E.M. Reproduction and obesity. Voeding, 16, 94-102, 1955.

Widdowson, E.M. Intra-uterine growth retardation in the pig. 1. Organ size and cellular development at birth and after growth to maturity. Biol. Neonate, 19, 329-340, 1971.

FACTORS INFLUENCING AMINO ACID UTILIZATION

Hamish N. Munro

Department of Nutrition and Food Science

Massachusetts Institute of Technology, Cambridge, Mass.

INTRODUCTION

Parenteral administration is an abnormal mode of alimentation. With regard to amino acid intake, the normal sequence of events involves metabolic changes in the intestine and screening of the incoming amino acids by the liver before they pass into the peripheral circulation. Once in the systemic plasma, the level of free amino acids is only moderately elevated after each protein meal of normal composition, thus implying that tissue removal of amino acids is adequate to prevent accumulation when the amino acid supply is delivered at a steady rate by way of the intestinal tract.

Accordingly, the questions which we must consider are the mechanisms involved at each of these stages in utilization of dietary amino acids and whether the delivery of amino acids parenterally has any potential or actual disadvantages due to the by-passing of these stages in normal assimilation. We shall therefore deal in turn with (a) the role of the digestive tract in response to a protein meal, (b) the role of the liver in amino acid metabolism, (c) the significance of changes in blood amino acid levels as indicators of amino acid utilization, (d) the role of skeletal muscle in amino acid utilization, and (e) the magnitude of amino acid turnover in relation to intake and requirements.

THE ROLE OF THE DIGESTIVE TRACT

It is now generally accepted that free amino acids without significant amounts of peptides are the digestive products of

proteins that pass into the portal blood. The dietary proteins are
only partly degraded by digestive enzymes within the gut lumen to
free amino acids, and the final resolution of small peptides to
amino acids is carried out by peptidases present in the mucosal
cells of the intestinal villi (Fauconneau and Michel, 1970). In
another respect, the intestinal mucosa is more than a transport
system, since it carries out transamination of glutamic and aspartic acids as they pass through the cell. In consequence, only
part of these dietary constituents are absorbed as such into the
portal blood (Elwyn, 1970). It is also probable that the liver
provides a second defence against elevated levels of absorbed glutamate, especially in the newborn animal (Stegink et al., 1971).
This implies that parenteral solutions containing large amounts of
glutamic or aspartic acids do not mimic the natural digestive fate
of proteins when their amino acids reach the systemic circulation.
Although this may suggest the danger of brain damage from the glutamate or aspartate in hydrolysates (Olney and Sharpe, 1969; Olney
and Ho, 1970; Olney, 1969), there is a report by Stegink and Baker
(1971) that two protein hydrolysates known to contain considerable
amounts of free glutamic and aspartic acids did not raise the blood
glutamate and aspartate levels of infants and did not cause obvious
neurological changes when given parenterally.

In addition to protein of dietary origin, the intestinal tract
also contains proteins secreted as enzymes as part of the digestive
function of the tract. A very considerable amount of protein is
also discharged into the tract by shedding of mucosal cells. This
endogenous protein appears to be added to the gut contents in
response to meals but is absorbed somewhat more slowly than dietary
protein (Gitler, 1964). The magnitude of this contribution is
quite considerable. Table 1 shows estimates of secreted protein in

Table 1. Daily Nitrogen Exchange in the Digestive Tract of a
70-kg Man. (From Fauconneau and Michel, 1970).

Component	Amount of Protein (gm/day)
Protein secreted into alimentary tract	
Saliva	3
Gastric juice	5
Bile	1
Pancreatic juice	8
Mucosal shedding	50
Balance sheet of protein entering and leaving tract	
Average dietary intake	90
Total secreted into tract	67
Fecal output	10
Amount absorbed (157-10)	147

comparison with dietary protein for a man of 70 kg body weight. Both dietary and endogenous protein appear to be efficiently absorbed, since fecal nitrogen losses are quite small.

THE ROLE OF THE LIVER

Although the dietary amino acids raise the levels of free amino acids in the portal vein considerably during the absorptive period, their concentrations in the systemic circulation are much less affected (Denton and Elvehjem, 1954), due to the capacity of the liver to remove amino acids from the portal blood and thus to regulate their flow into the systemic circulation. This is described by Elwyn (1970) who fed a meal of meat to dogs and found that 57% of the absorbed amino-N was transformed to urea-N on one passage through the liver, while only 23% passed into the general circulation as free amino acids, 6% as plasma proteins, and the remaining 14% was retained in the liver transiently as protein. It can thus be said that the amino acid nutrition of the body is determined by the liver.

This dramatic intervention of the liver in amino acid metabolism after meals causes diurnal rhythms in liver protein metabolism, which affects both rate of synthesis of liver proteins and also the levels of enzymes of amino acid catabolism (Fishman et al., 1969). The influx of amino acids after each meal leads to a temporary increase in the abundance of polysomes actively synthesizing proteins. Some of the proteins so synthesized are enzymes of amino acid catabolism, which also increase in amount because they are stabilized by their substrates. Consequently, when rats are fed ad libitum, parallel diurnal rhythms in their liver polysomes and in the liver catabolic enzyme tyrosine aminotransferase are observed (Fishman et al., 1969). More detailed examination of the metabolic changes occurring in the liver after a meal of protein shows that regulatory responses occur in many aspects of liver cell function. Thus, while the aggregation of polysomes indicates increased protein synthesis, there is an accompanying reduction in liver RNA breakdown and an increased de novo biosynthesis of purines by the liver (Fig. 1).

As a result of these meal-related adaptative liver responses, the systemic circulation is protected against excessive changes in free amino acid concentration entering the body. Elwyn's results were obtained following a large protein meal. Other studies suggest that these proportions would be different if a diet lower in protein had been fed, and that a larger fraction of the incoming amino acids would pass into the general circulation, notably the essential amino acids. Thus many of the enzymes of amino acid metabolism undergo adaptive changes related to alterations in amino acid

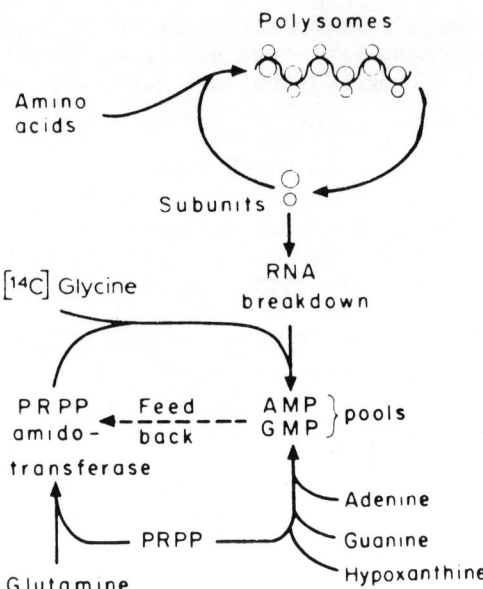

Fig. 1. Schematic representation of the effect of amino acid supply to the liver on protein synthesis by liver polysomes, and on RNA degradation rate and synthesis of purine nucleotides. The diagram indicates interrelationships between those metabolic events which result in reduced RNA breakdown and increased purine biosynthesis when amino acid supply to the liver is increased (Clifford et al., 1972).

supply (see Kaplan and Pitot, 1970, for review). Harper (1968) has examined how enzymes degrading <u>essential</u> amino acids behave when the intake of a given amino acid is below needs or when the diet provides an excess of the amino acid over requirements. Young rats were given different levels of dietary casein from insufficient up to excessive quantities. Threonineserine dehydratase activity in the liver remained low until the casein content of the diet reached 20% which is optimal for growth of the rat. At casein intakes above 20%, enzyme activity rose sharply. On the other hand, a transaminase handling the non-essential amino acid, glutamic acid, increased linearly in proportion to the intake of casein.

These observations predict that dietary essential amino acids will not be extensively degraded in the body until they are eaten in excessive quantities. Brookes et al. (1972) have recently tested this hypothesis for lysine fed to rats. Increasing amounts of ^{14}C-lysine were added to the diet of the growing rat, other dietary amino acids being kept constant. Consumption of amounts of lysine below requirements led to somewhat similar small proportions of the total dietary ^{14}C-lysine being released as ^{14}CO$_2$. However, when lysine intake was raised above the requirement for optimal growth, the proportion of ^{14}C-lysine released as ^{14}CO$_2$ rose sharply and further addition of lysine to the diet increased ^{14}CO$_2$ production still further. There was a sharp point of inflection of ^{14}CO$_2$ output at the point where lysine intake was just sufficient to support the maximal growth rate.

SYSTEMIC BLOOD AMINO ACID LEVELS AND AMINO ACID UTILIZATION

The use of plasma amino acid levels as an index of nutritive status has been extensively explored (see review by Munro, 1970). Characteristic changes are seen in established protein malnutrition, there being a fall in the levels of most essential amino acids, combined with a rise in the levels of several non-essential amino acids.

A potentially very useful feature of systemic amino acid levels is their response to increasing levels of one essential amino acid. Several investigators (Zimmerman and Scott, 1965; Mitchell et al., 1968; Pawlak and Pion, 1968; Stockland et al., 1970, 1971) have examined the effect by adding increasing amounts of the limiting essential amino acid to a deficient diet and observing the changes occurring in the free amino acid levels in the blood as the intake of the limiting amino acid progressed from insufficient to excess. For example, in the case of lysine, which is required by growing chicks, it was found that the plasma level of this amino acid did not rise appreciably until the lysine intake represented 0.8% of the diet, after which there was a sharp rise in plasma lysine concentration (Zimmerman and Scott, 1965). Maximal growth rate was also attained when the lysine content of the diet reached 0.8%. Thus the dietary concentration of lysine at which the plasma level of that amino acid begins to rise sharply is an indicator of adequacy of lysine intake. These authors made similar observations with arginine and valine.

This approach has also been used to study the amino acid requirements of human subjects. Young et al. (1971) have recently examined the effect of giving increasing levels of dietary tryptophan on the free tryptophan content of plasma and the N balance of young adults (Fig. 2). Their study confirmed the relationship between adequacy of essential amino acid intake and blood levels of the amino acid. At intakes of tryptophan below 3 mg per kg, the plasma levels of tryptophan were low and constant. Between intakes of 3 and 5 mg, tryptophan concentration rose sharply but

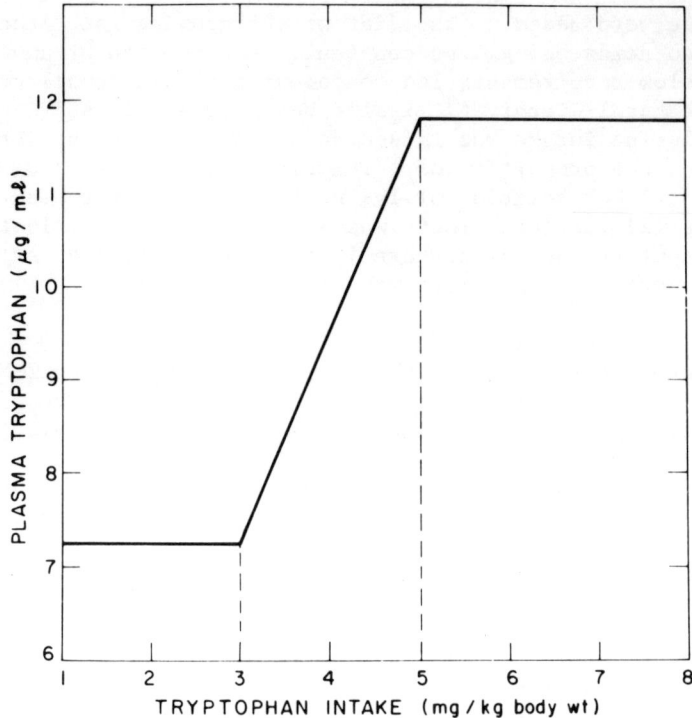

Fig. 2. Schematic representation of plasma free tryptophan responses to graded levels of dietary tryptophan fed to young men receiving a constant amount of all other amino acids. (Drawn from the data of Young et al., 1971).

at intakes beyond 5 mg it reached a second plateau at still higher level. The sharp increment at 3 mg intake was taken to mean that the requirements of the subjects had been met at this point; the second plateau was interpreted as induction of tryptophan pyrrolase in the liver. Nitrogen balance measurements showed that these young adults attained N equilibrium between 2 and 2.6 mg tryptophan per kg; however, the authors point out that no allowance was made for integumental and sweat losses in measuring N balance, and in all probability the true N equilibrium was not attained until an intake of about 3 mg tryptophan per kg body weight had been reached. These studies consequently support the use of blood amino acid measurement as a means of determining the amino acid needs of man and other animals, and an alternative to the need to do N balances.

In a recent study (Young and Munro, 1971), we have examined the effect of age of the rat on the response of its plasma tryptophan concentration to different levels of dietary tryptophan. The

FACTORS INFLUENCING AMINO ACID UTILIZATION 17

rats used were either very young (54 g weight initially) or mature but not very old adults (320 g weight). Although the findings do not apply to rats in old age, they are nevertheless interesting in showing that the method is sensitive to age-related changes in requirements. Rats from each of the two age-groups were given diets providing amino acids in place of protein, and the intake of tryptophan was varied from zero up to 0.33% of the diet. The requirement for tryptophan in the young has been given by Rama Rao et al. (1959) as 0.11% and for mature rats by Smith and Johnson (1968) as 0.03%. These requirements are indicated in Fig. 3 by arrows. Plasma tryptophan was measured twice a day (at 11 a.m. and

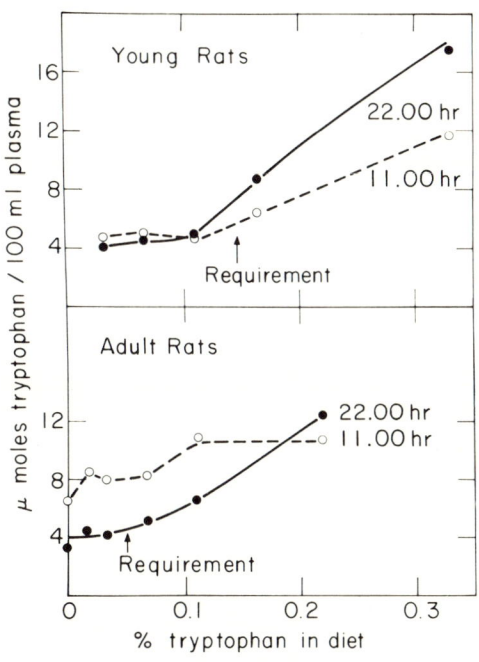

Fig. 3. Plasma tryptophan levels of young and mature rats fed amino acid diets containing various levels of tryptophan (Young and Munro, 1971).

11 p.m.) following a 9-day period on these diets. Fig. 3 shows that the plasma tryptophan increases at both these times in the case of the young rats as soon as the requirement has been met. In the case of the mature animals, the increase is evident for blood sampled at 11 p.m., which coincides with the rats habitual time of feeding, whereas the results obtained at 11 a.m. do not provide this clear evidence of a requirement-dependent increment. This indicates that the use of blood amino acid responses for determining requirements may be more sensitive during the absorption phase. From these studies, it would appear to be feasible to examine parenteral amino acid requirements by observing the response of plasma amino acid levels to increasing amounts of one amino acid in the infusion. It will be apparent that such methods can also be used to assess the amino acid needs of patients with various diseases.

Intake of protein is not the only dietary factor causing changes in plasma free amino acid levels. The free amino acids of the plasma are also depressed by administering carbohydrate. This effect starts shortly after giving glucose and appears to be due to insulin-dependent deposition of amino acids in skeletal muscle (Munro, 1964). This deposition results in a diminished amino acid supply to the liver and in consequence less urea production, the so-called protein sparing action of carbohydrate. The combination of carbohydrate and protein in the same meal seems to result in an even greater specific protein-sparing action due to the same mechanism. Cuthbertson and Munro (1939) showed that consumption of the protein and the carbohydrate of a diet in separate meals results in a temporary phase of negative nitrogen balance which is corrected by taking the same amounts of carbohydrate and protein in mixed meals. A similar phenomenon has been recorded for parenteral nutrition (McNair, 1960). This suggests that the giving of carbohydrate along with amino acids has a beneficial effect on the metabolism of the amino acids, more of which are deposited in muscle under the action of the carbohydrate.

THE ROLE OF SKELETAL MUSCLE

Skeletal muscle accounts for 45% of the weight of the non-obese adult, and acts as a depository of half or more of the free amino acids present in the body (Munro, 1970). Furthermore, transport of amino acids into muscle and subsequent deposition as muscle protein are under the control of insulin (Munro, 1970). It can be concluded that removal of infused amino acids is to some extent dependent on skeletal muscle. However, in the case of the newborn infant, skeletal muscle accounts for only about 25% of body weight (Miller, 1969) and this must alter the proportion of infused amino acids passing into the musculature.

The role of skeletal muscle in the amino metabolism of man has recently been elegantly elucidated by Cahill and his colleagues (Pozefsky et al., 1969; Felig and Wahren, 1971). By measuring arterio-venous differences in amino acid concentration in the forearm or legs of human subjects, they were able to quantitate the amounts of individual amino acids added to or removed from the limb muscles as the blood passed through. Table 2 shows that after an overnight fast there is a net output of amino acids from the musculature equivalent to a daily loss of 75 gm of muscle protein.

Table 2. Concentration of Amino Acids in Muscle Protein Compared with the A-V Plasma Difference Across the Arm and Leg of Fasting and Insulin-Injected Subjects.

Measurement on	Total amino acids	Lysine	Alanine
Mixed muscle protein	100%	10%	6%
A-V difference in fasting (μmoles/L. plasma):			
A.A. release by arm	490 (100%)	44 (9%)	120 (25%)
A.A. release by leg	160 (100%)	19 (12%)	68 (42%)
Reduction in A-V arm by insulin (μmoles/L. plasma):			
Control (fasting)	570	44	118
Insulin 1 hour	150	33	98
Diff. due to insulin	-420 (100%)	-11 (2%)	-20 (5%)

Calculated from Felig and Wahren (1971) and from Pozefsky et al. (1969) who estimate 0.43 moles amino acids (=50 gm protein) released daily from musculature of post-absorptive man. With inclusion of glutamine N release (not shown in table) muscle protein catabolism is probably 75 gm daily.

However, the table shows that, in contrast to the relative concentrations of amino acids found in muscle protein, a large part of this released amino acid N is made up of alanine and glutamine (not shown in the table). It can be concluded that much of the amino acid pool in muscle becomes transaminated to these two amino acids. These two amino acids released from muscle are then quantitatively taken up by the liver where the alanine provides a source of carbon for gluconeogenesis and both alanine and glutamine are donors of NH_2 for urea synthesis (Fig. 4). When insulin

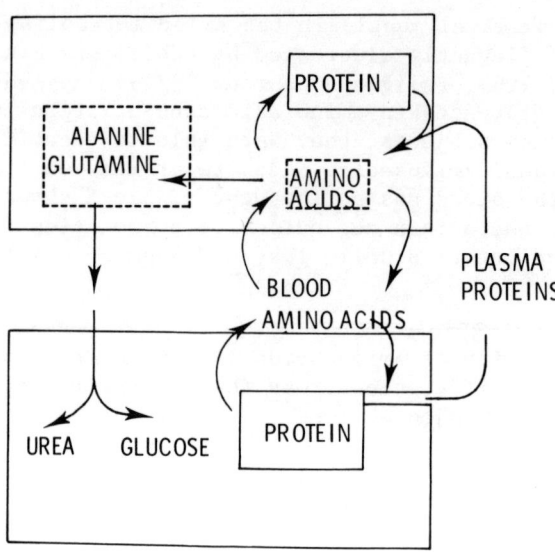

Fig. 4. Diagram showing recycling of amino acids within skeletal muscle and between muscle and liver, mainly as alanine and glutamine.

is injected into fasting subjects, this release of amino N into the blood-stream is sharply reduced (Table 2); this phenomenon has been considered to be due to a reduction in rate of degradation of muscle proteins (Pozefsky et al., 1969). However, this could also occur if the well-known stimulation of muscle protein synthesis by insulin results in more rapid removal of amino acids from the intracellular pool, so that less is available for release into the blood. These alternatives could be directly tested if an amino acid is released from muscle breakdown that is not reutilized for protein synthesis. If muscle protein breakdown is indeed diminished by insulin, then output of such a non-reutilizable amino acid should diminish. In fact, we appear to have identified such an amino acid in the form of 3-methylhistidine (Young et al., 1972). This is present in both the actin and myosin of muscle and is synthesized by methylation of histidine after histidine has been incorporated in the peptide chains during synthesis of these proteins (Fig. 5). Study of the fate of injected 3-methylhistidine shows that in part, it is metabolized by the rat to form N-acetyl-3-methylhistidine, and that the free and conjugated 3-methylhistidine are rapidly excreted in the urine in quantities equivalent to 100% of the original dose administered. Thus, the release of 3-methylhistidine

FACTORS INFLUENCING AMINO ACID UTILIZATION 21

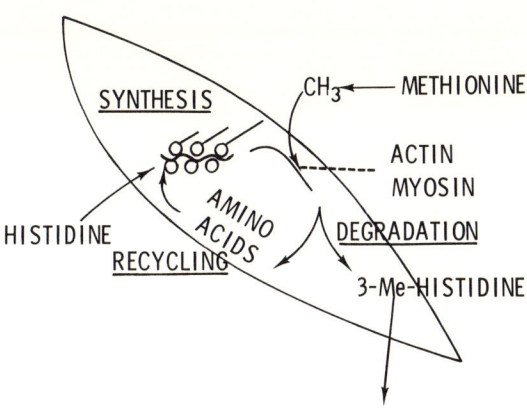

Fig. 5. Diagram to illustrate the use of 3-methylhistidine in measuring rate of muscle protein breakdown.

from muscle into the blood and its excretion in urine may provide an absolute measure of the rate of muscle protein degradation. However, it will be necessary to exclude significant contributions of 3-methylhistidine from other tissues before this technique can be finalized as suitable for measuring muscle protein breakdown on the basis of urinary output.

THE DAILY FLUX OF AMINO ACIDS IN THE BODY

From the information on amino acid metabolism already considered, together with some additional data, we can begin to assemble a composite picture of the daily flux of amino acids in various compartments of the body of an adult man. As applied to infants, such figures would, of course, require modification not only as regards total amounts but also in their relative magnitudes in different tissues. For example, the proportion of muscle on the infant body is less than in the adult. The best estimates of the daily amino acid flux in different compartments of the body of a 70 kg man are shown in Fig. 6.

Some comments on the data are needed. Balance experiments show (Munro, 1972) that the <u>average</u> man of this size can be maintained in nitrogen equilibrium on an intake of 32 gm of high quality dietary protein. However, the customary protein intake in Western countries is about 90 gm daily. This is augmented by addition of some 70 gm of protein secreted into the gastro-intestinal tract (Table 1), so that the total load for adsorption is about 150 gm. Although this is a large amount, the pools of free amino acids in

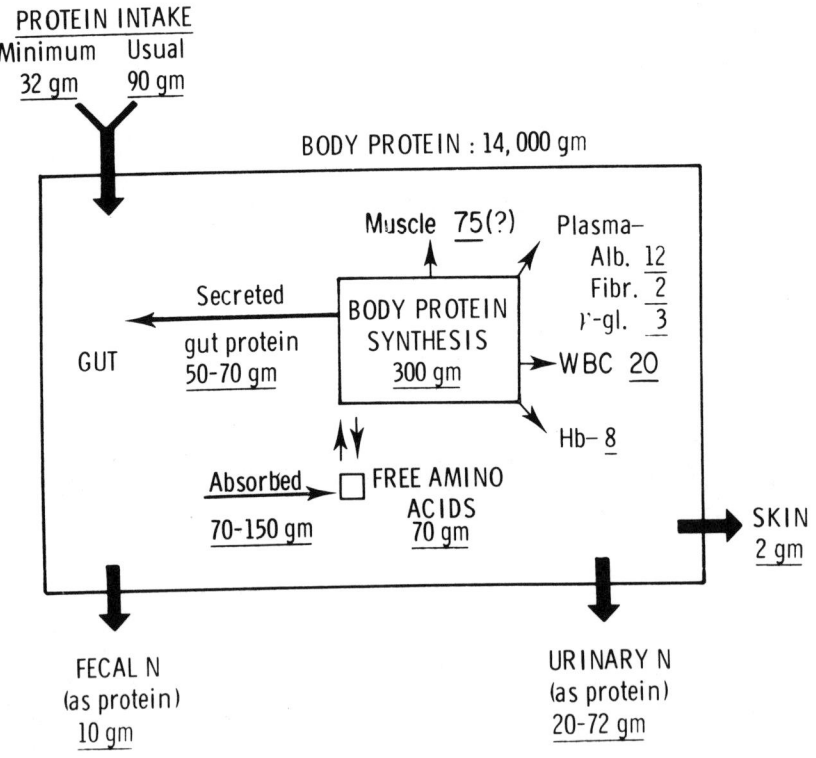

Fig. 6. Diagram to show daily amino acid flux in the body of a 70 kg man.

Table 3. Intake, Tissue Content and Turnover of Essential and Non-Essential Amino Acids for a 70 kg Adult.

Amino acid source	Amino Acids per 70 kg B. Wt.		
	Total	Essential	Non-Essential
	gm	gm	gm
Daily diet			
Minimum amino acid needs	32	6	26
Western diet (1)	90	45	45
Absorbed (with secreted gut protein) (1)	150	75	75
Free Amino acid pools (2)			
Plasma	0.7	0.2	0.5
Tissues	70	10	60
Daily body protein turnover (3)	300	150	150

(1) These data are taken from Table 1. It is assumed that the mixed proteins of the diet and the intestinal secretions contain approximately 50% essential amino acids.

(2) These are calculated for a 70 kg adult from data on free amino acid concentrations in the blood and tissues of the rat (Munro, 1970).

(3) From Fauconneau and Michel (1970), assuming that 50% of the amino acids turned over are essential amino acids.

the tissues are about 70 gm and could thus accommodate the incoming amino acid load without serious expansion, even if metabolic removal did not occur. The free amino acid pool exchanges with body protein and experiments involving ^{15}N suggest that about 300 gm protein is synthesized daily by the body of an adult. Synthesis rates for the proteins in some specific tissues are shown in Fig. 6, and account for two-thirds of the 300 gm daily. Nevertheless, the data shown in Fig. 6 should only be regarded as provisional.

In Table 3, some of these components in daily amino acid flux are expressed as amounts of essential and non-essential amino acids.

The average minimum requirement of dietary protein for N equilibrium in the adult need only contain 6 gm of essential amino acids. However, the diet of 90 gm of protein usually consumed provides about 45 gm of essential amino acids. To this are added the essential amino acids entering the gut as endogenous proteins. In consequence, about 75 gm of essential amino acids are absorbed daily. This amount can be contrasted with the total amounts of essential and non-essential amino acids in the plasma, namely 0.2 and 0.5 gm. It is thus to be expected that free amino acids will be rapidly transported out of the blood into the tissues. It will also be noted that, although the tissues contain some 70 gm of free amino acids, only 10 gm are as essential amino acids. Nevertheless, the daily turnover of 300 gm of body protein requires 150 gm of essential amino acids. Most of this has to come from recycling of essential amino acid released by the tissues, since the diet customarily provides about 45 gm essential amino acids. Nitrogen equilibrium can still be achieved when the essential amino acid content of the diet is as low as 6 gm. This implies very efficient recycling of amino acids within the body.

For the future, we need more accurate estimates of amino acid fluxes in different tissues, not only in healthy adult subjects but also in patients with injuries or debilitating diseases, and also for children at various stages of development. Such information would allow better prediction of the needs of various groups of patients for amino acids given parenterally.

REFERENCES

Brookes, I.M., Owens, F.N. and Farrigus, U.S., J. Nutr., 102, 27, 1972.

Clifford, A.J., Riumallo, J.A., Baliga, B.S., Munro, H.N. and Brown, P.R., Biochim. Biophys. Acta, 1972 (in press).

Cuthbertson, D.P. and Munro, H.N., Biochem. J., 33, 128, 1939.

Denton, A.E. and Elvehjem, C.A., J. Biol. Chem., 206, 449, 1954.

Elwyn, D., in Mammalian Protein Metabolism (ed. H.N. Munro), vol. IV, Ch. 38, p. 523, Academic Press, New York, 1970.

Fauconneau, G. and Michel, M.C., in Mammalian Protein Metabolism (ed. H.N. Munro), vol. IV, Ch. 37, p. 481, Academic Press, New York, 1970.

Felig, P. and Wahren, J., J. Clin. Invest., 50, 2703, 1971.

Fishman, B., Wurtman, R.J. and Munro, H.N., Proc. Nat. Acad. Sci., 64, 677, 1969.

Gitler, C., in Mammlian Protein Metabolism (ed. H.N. Munro and J.B. Allison), vol. I, Ch. 2, p. 35, Academic Press, New York, 1964.

Harper, A.E., Amer. J. Clin. Nutr., 21, 358, 1968.

Kaplan, J.H. and Pitot, H.C., in Mammalian Protein Metabolism (ed. H.N. Munro), vol. IV, Ch. 35, p. 387, Academic Press, New York, 1970.

McNair, R.D., Clin. Chem., 6, 115, 1960.

Miller, S.A., in Mammalian Protein Metabolism (ed. H.N. Munro), vol. III, Ch. 26, p. 183, Academic Press, New York, 1969.

Mitchell, J.R., Becker, D.E., Jensen, A.H., Harmon, B.G. and Norton, H.W., J. Animal Sci., 27, 1327, 1968.

Munro, H.N., in Mammalian Protein Metabolism (ed. H.N. Munro and J.B. Allison), vol. I, Ch. 10, p. 381, Academic Press, New York, 1964.

Munro, H.N., in Mammalian Protein Metabolism (ed. H.N. Munro), vol. IV, Ch. 34, p. 299, Academic Press, New York, 1970.

Munro, H.N., in Parenteral Nutrition (ed. A. Wilkinson), Balliere, London (in press), 1972.

Pawlak, M. and Pion, M., Annal. Biol. Animal Biochem. Biophys., 8, 517, 1968.

Pozefsky, T., Felig, P., Tobin, J., Soeldner, J.S. and Cahill, G.F., J. Clin. Invest., 48, 2273, 1969.

Olney, J.W., J. Neuropath. Exp. Neurol., 28, 455, 1969.

Olney, J.W. and Sharpe, L.G., Science, 166, 386, 1969.

Olney, J.W. and Ho, O.L., Nature, 227, 609, 1970.

Rama Rao, P.B., Metta, V.C. and Johnson, B.C., J. Nutr., 69, 387, 1959.

Smith, E.B. and Johnson, B.C., Brit. J. Nutr., 21, 17, 1967.

Stegink, L.D. and Baker, G.L., J. Ped., 78, 595, 1971.

Stegink, L.D., Baker, G.L. and Filer, L.J., cited by Stegink and Baker, 1971.

Stockland, W.L., Meade, R.J. and Melliere, A.L., J. Nutr., 100, 925, 1970.

Stockland, W.L., Lai, Y.F., Meade, R.J., Sowers, J.E. and Oestemer, G., J. Nutr., 101, 177, 1971.

Young, V.R., Mohammed, A., Hussein, E.M. and Scrimshaw, N.S., J. Nutrition, 101, 45, 1971.

Young, V.R. and Munro, H.N., Unpublished studies, 1971.

Young, V.R., Alexis, S.D., Baliga, B.S., Munro, H.N. and Muecke, W., J. Biol. Chem., 247, 3592, 1972.

Zimmerman, R.A. and Scott, H.N., J. Nutr., 87, 13, 1965.

AMINO ACID REQUIREMENTS IN CHILDHOOD

D. M. Hegsted

Department of Nutrition, Harvard School of Public

Health, Boston, Massachusetts

The sources of data upon the amino acid requirements of infants and children are very limited. The major work is that of Holt and Snyderman and their associates and their estimates have been published in many places. When data are limited there is also a limit to what one can say about them. However, one of the major problems is that people are likely to take estimates of requirements too seriously without adequate consideration of the limitations of the data.

The World Health Organization and the Food and Agriculture Organization convened another Expert Committee nearly two years ago to re-evaluate protein and energy requirements. This report is not yet available. Differences in philosophy, differences in experimental results, differences in interpretation of experimental results, and often very limited data to deal with made it very difficult for this group to arrive at a consensus. As in many areas of research, the more one knows the more evident it is how little is actually known. Research tends to uncover more problems than it solves and it is almost certainly true that the general area of protein and amino acid requirements is more controversial now than it was a few years ago. The ultimate solution to these disagreements must be more and better work--they are not likely to be solved by committees or argument although these are often useful.

In estimating the amino acid requirements of infants, there are essentially two sources of data. These are the experiments of Holt and Snyderman (1967) who fed diets in which essentially all of the protein was replaced by a mixture of amino acids. The concentration of each essential amino acid could be varied at will and they measured growth and nitrogen retention at various levels of intake of each essential amino acid. A relatively small number of infants could be studied in these rather tedious experiments and the estimated requirement (the smallest amount judged to give maximum growth and nitrogen retention) was found to vary substantially from one infant to another. Thus, they selected the highest value found as the estimate of the requirement. As Fomon and Filer (1967) have pointed out, these kinds of studies may be criticized as follows: a) mixtures of amino acids may not be nutritionally equivalent to whole proteins; b) the studies have been done with relatively high intakes of total nitrogen (equivalent to about 3 gm of protein per kg per day) and the intake of total nitrogen might influence amino acid needs; c) few subjects were studied for each amino acid; d) each study was of relatively short duration; e) the criteria, nitrogen balance and growth, may not be sufficiently sensitive especially in short experiments. These kinds of criticisms only indicate the reservations that one must have about the final values obtained. Certainly, the studies of Holt and Snyderman are classic.

Table I: Range of Estimated Requirements, mg/kg/day.

	Holt and Snyderman	Fomon and Filer
Histidine	16 - 34	19 - 28
Isoleucine	102 - 119	48 - 70
Leucine	76 - 229	111 - 161
Lysine	88 - 103	111 - 161
Methionine*	33 - 45	23 - 29
Phenylalanine[+]	47 - 90	42 - 61
Threonine	45 - 87	80 - 116
Tryptophan	15 - 22	11 - 17
Valine	85 - 105	64 - 93

*Methionine requirement in the presence of crystine.
[+]Phenylalanine requirement in the presence of tyrosine.

The other kind of data that is available is the kind presented by Fomon and Filer (1967) in which infants have been fed various diets which do allow normal growth and development and the amino acid content of these diets examined. They fed 22 infants a milk-based formula containing 6% of the calories as protein and studied their performance and intake. Nineteen were judged to have developed normally. The amino acid intakes of these were presented. It would seem clear that if an infant develops normally over an extended period, the intake can be judged to have been adequate. On the other hand, such data do not indicate a requirement. They only indicate that the intake was at or above requirement.

The FAO/WHO Committee examined the highest estimates of Holt and Snyderman and those of Fomon and Filer as shown in Table 1. They assumed that the requirements cannot be above those estimated by Fomon and Filer but might be below, and thus the lower of the two estimates might be the more logical estimate of the real requirement. In the last column of Table 2 are shown the concentrations of amino acids in dietary protein which would be required assuming that the infant needs 2 gm of protein per kg per day.

Table 2

Estimated Amino Acid Requirements of Infants

Amino Acids	Estimated Requirements		Composite of Lower Values	Suggested Pattern[3]
	Holt & Snyderman[1]	Fomon & Filer[2]		
	mg/kg/day	mg/kg/day	mg/kg/day	mg/g protein
Histidine	34	28	28	14
Isoleucine	119	70	70	35
Leucine	229	161	161	80
Lysine	103	161	103	52
Methionine + cystine	45 + cys	58[4]	58	29
Phenylalanine + tyrosine	90 + try	125[4]	125	63
Threonine	87	116	87	44
Tryptophan	22	17	17	8.5
Valine	105	93	93	47

[1] Requirements estimated when amino acids were fed or incorporated in basal formulas. The values represent estimates of maximal individual requirements to achieve normal growth (1).
[2] Calculated intakes of amino acids when formulas were fed in amounts sufficient to maintain good growth in all the infants studies--the amino acids were not varied independently (2).
[3] Based on a safe level of intake of 2 g protein/kg/day, the average of suggested levels for the period 0-6 months.
[4] The values for cystine and tyrosine were estimated on the basis of the methionine:cystine and phenylalanine:tyrosine ratios in human milk.

It is necessary to emphasize the difficulties. Obviously, clinical judgment is an important criteria and cannot be easily quantitated. Nitrogen balance, if it is well done, tells us what goes in and what comes out. It does not tell us what goes on inside the body. It is at least possible that specific tissues or biochemical systems might be depleted or modified adversely by diets which do not affect overall nitrogen balance. One becomes increasingly convinced that nitrogen balance is not a very sensitive tool under many conditions. Graham (1967) has presented data using partially depleted infants recovering from malnutrition in which there is a somewhat inverse relationship between gain in weight and regeneration of serum albumin. The body apparently makes some choices as to what it does with a limited protein intake but what determines these choices is not known. However one looks at the data available, it would seem quite evident that much more sensitive tools would be very useful if these can be found.

All of the data indicate that some infants do well upon diets which are inadequate for others. The estimates of need may vary nearly 100% in some instances. It is practically certain that some of this variation is due to the fact that our tools are not very sharp, so a certain but unknown proportion of the discrepancy must be due to error. However, it is most unlikely that all of the variation can be explained in this way and we must conclude that genetic or other factors have very significant effects on nutritional requirements.

The published requirements thus represent estimates of the upper limits of requirements. The reason for this is obvious. It should also be clear that a diagnosis of deficiency cannot be made when intake falls below these estimates unless it is far below. The only thing one can say is that the "risk of defiency" increases as the intake falls below the estimated need.

When we turn to estimates of the amino acid needs of children we are in a very bad situation indeed. There are almost no data which are directly relevant to the question except for the observations of Nakagawa et al. (1961-1963) on school children 10 to 12 years of age. The studies of Nakagawa et al. were also done with few subjects fed diets containing amino acids. Requirements were estimated from nitrogen balance and the highest observed value was selected.

The general approach which has been used to estimate the protein needs of children has been to consider them as small adults which grown at different rates. That is, the maintenance requirement has been estimated based upon body weight, assuming that their requirement is governed by the same parameters as in adults (basal metabolic rate) and to this value is added a requirement based upon the amount of tissue protein that needs to be deposited to allow for a normal rate of growth.

Table 3: Estimated Amino Acid Requirements of Children

Amino Acid	School children, 10-12 years	
	Observed Requirement[1]	Suggested Pattern[2]
	mg/kg/day	mg/g protein
Histidine	0	0
Isoleucine	30	37
Leucine	45	56
Lysine	60	75
Methionine + cystine	27	34
Phenylalanine + tyrosine	27	34
Threonine	35	44
Tryptophan	4	4.6
Valine	33	41

[1]Based on Nakagawa et al. (1961-1963). The values represent estimates of the upper range of individual requirements for the achievement of positive nitrogen balance in boys.

[2]Based on a safe level of protein intake of 0.8 g/kg/day, the average of safe levels of protein for boys and girls in that age-group.

The legitimacy of this approach has never been adequately tested. Although we cannot doubt that growth influences protein and amino acid requirements, the primary questions are whether the relative proportions of amino acids which are needed are similar for growth and maintenance. They probably are not. Secondly, what efficiency of utilization can one expect in the deposition of the body protein.

All of the data available suggest that total protein needs fall rapidly during the first few months of life. Unfortunately, the curves which have been presented are based largely upon rates of growth rather than quantitative estimates, but there is little doubt that they fall. Practically no data are available upon amino acid needs although Holt and Snyderman (1967) summarized data upon the phenylalanine requirement of phenylketonurics of different ages and suggest that there is a rapid fall from an average of about 80 mg per kg per day at birth (a value consistent with Holt's estimate of 90 mg per kg per day for normal infants) to about 35 mg per kg per day at 12 months of age (Fig. 1).

The latter value is not much above the estimate of 27 mg per kg per day for phenylalanine plus tyrosine obtained by Nakagawa et al. (1961-1963). The latter values as summarized by the FAO/WHO Committee are shown in Table 2 together with the concentration required in the dietary protein if the requirement is 0.8 gm per kg per day.

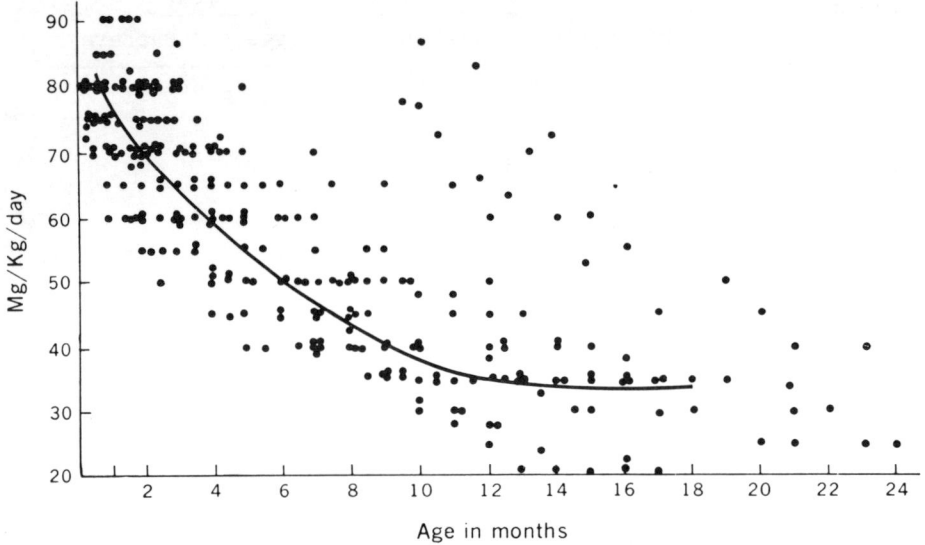

Fig. 1. Relationship of phenylalanine requirement to age in 27 phenylketonuric patients (Holt & Snyderman, 1967).

The estimates of the requirements for essential amino acids in adults are lower relative to the total nitrogen requirements than in infants and children, i.e., the requirements for essentail amino acids fall faster or to a greater extent with age than do the total nitrogen needs. The total protein requirement is estimated to be about 2 gm per kg per day for the infant and about 0.55 gm per kg per day for the adult--about 25% of the infant value. On the other hand, the fall in the estimated requirement of the essential amino acids is nearly 10-fold. For example, the estimated requirement of lysine for the infant is 103 mg per kg per day and for the adult only 12, or the estimated requirement of threonine is 87 mg per kg per day for the infant and only 7 for the adult.

No one has an adequate explanation for this. Is it due to the difference in methodology of estimating requirements? Nitrogen balance and growth are more sensitive indicators in infants than is nitrogen balance alone in the adult. Is growth in the human infant a very inefficient process compared to maintenance in the adult as far as essential amino acids are concerned? Are there basic differences in the metabolism of the growing and adult animal so that one cannot consider the growth requirement as a simple addition to the maintenance need?

A 10 to 12 year old child will grow about 4 kg per year or about 10 gm per day on the average and weighs 30-40 kg. This should mean a deposition of about 2 gm of protein as tissue per day and this seems relatively insignificant compared to a maintenance requirement of 30-40 gm of protein per day. It is difficult to square Nakagawa et al.'s values of amino acid needs which are 2 to 3 times those of the adult with this relatively slow rate of growth. Nakagawa et al.'s estimates may be high, of course, but the data do suggest that there may be more fundamental differences in the metabolic process and conversions of amino acids to tissue protein than has generally been thought and that we cannot consider the child as a growing adult.

Values for amino acid needs between those of infants and the 10 and 12 year old are non-existant. Similarly, there are no direct data between the 10 to 12 year old child and adults. One can only attempt to draw curves between these points. Obviously, an unsatisfactory situation.

A few other unsolved or controversial points should be mentioned. Some proteins appear to be unsatisfactory for young children as a sole source of protein (Graham, 1967; Graham et al., 1966; Srikantia and Gopalan, 1966) even though their amino acid composition is satisfactory and they are of high quality by rat tests. As already mentioned, the total nitrogen intake probably influences the amino acid requirements and thus may make estimates under one set of conditions generally not applicable, although the magnitude of the effect is not known. Amino acid imbalances obviously affect requirements. These are probably of minor importance when proteins are fed but may need consideration when amino acid mixtures are being concocted. The composition of the non-essential nitrogen part of a mixture may be of significance. Some reports have indicated that calorie requirements are higher when amino acids are fed (Rose et al., 1954), others that they are not (Metta et al., 1960; Ahrens et al., 1966). This may reflect the composition of the non-essential nitrogen fraction. Finally, the whole matter of calorie intake and its effect upon amino acid and protein requirements is still not well studied. Utilization of amino acids and proteins are clearly compromised when intakes are low but there is disagreement on the extent of this effect at different intakes. Children recovering from protein-calorie malnutrition (Ashworth, 1970) apparently have very

high calorie requirements which are not understood and the same
may be true of other patients. The assumption made about the
calorie need may determine whether or not a mixture is adequate
or not.

Some of the difficulties in obtaining and interpreting data
in this field are inherent in the use of human subjects. Ethical
considerations place many restraints on what can be done. In
particular, one cannot produce severe deficiencies deliberately to
study the pathology and biochemical abnormalities of uncomplicated
deficiencies nor can one justify long-term studies. Considerable
assistance, both in obtaining data and in evaluating the data
currently available, might be obtained if we had better comparative
data with other species. Data upon amino acid needs are largely
limited to those obtained with young rats and chicks. The chick
clearly has limitations as a model for man. The rat has been
considered to be much more satisfactory but doubts are being raised.
In particular, the young rat seems to be a much more efficient
utilizer of protein than the infant which raises doubts about its
suitability as a model. The quantitative amino acid needs, as
we understand them today, certainly show similarities to man but
it is not likely that they are identical.

In our laboratory we have recently turned to studies with
young monkeys. The general approach has been similar to that used
by Holt with his infants, starting baby monkeys at eight weeks of
age and feeding diets with different levels of protein for a test
period of three weeks. An example of the data obtained is shown
in Fig. 2 in which the dietary protein was lactalbumin (Samonds
and Hegsted, in press). The animals weighed from 400 to 1,000 gm
and were fed arbitrarily selected levels of protein which ranged
over those levels which produced loss of weight to maximum gains.

As would be expected, small animals required less protein than
large animals. The regression of the weight gain to protein intake
for all of the data obtained and when the animals are divided into
three weight classes is shown in Fig. 2. The usual transformation
when dealing with animals or subjects of different size, of course,
is to correct for size by converting intake into mg per kg body
weight. When this is done, however, (Fig. 3) the results were
somewhat unexpected, but seem readily explainable.

The estimated requirement per unit weight for maintenance (zero
growth) appears to be independent of size. However, the small
animals have substantially larger requirements per unit body weight
than large animals to produce the same rate of gain. This would
appear to be due to the fact than an animal of 1,000 gm which receives 1 gm per kg of protein above the maintenance need has 1 gm

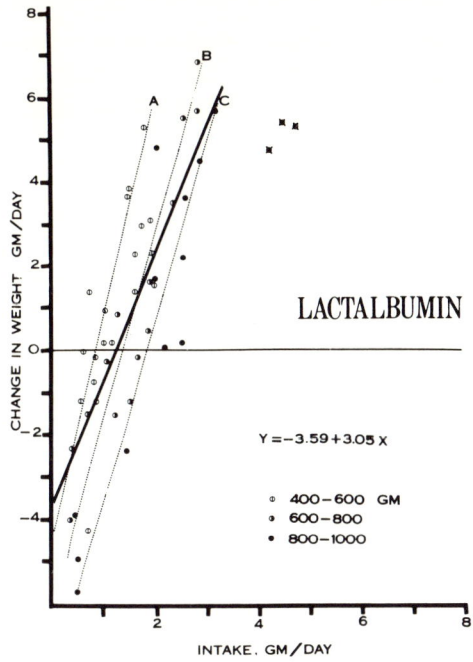

Fig. 2. Growth response of infant cebus monkeys fed different amounts of lactalbumin.

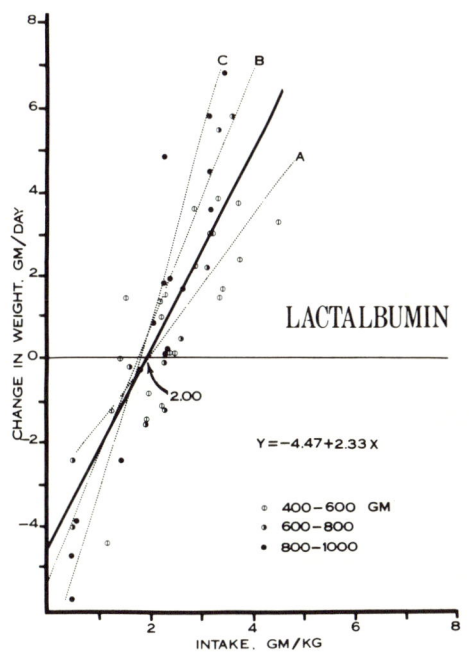

Fig. 3. Growth response of infant cebus monkeys to different amounts of lactalbumin (data from Fig. 1) expressed as intake per unit weight.

of protein available for the formation of new tissue. However, an animal of 500 gm weight which also receives 1 gm per kg above the maintenance need only gets 0.5 gm of protein which is available for new tissue formation. Thus, he is able to grow at only about half the rate.

We also estimate from these data that the growing monkey is only able to convert about 70% of the lactalbumin protein to new tissue protein. The same kind of data obtained with young growing rats yields values above 90% efficiency. Our value of about 70%

efficiency in the monkey is similar to the estimated efficiency of about 65% for egg protein in adult men and women studied by (Calloway and Margen (1971). This also suggests that the young rat may not be as satisfactory a model as is often supposed and that the sub-human primate may be better.

In summary, it should be clear that the published estimates of the amino acid requirements for infants represent an estimate of the amount which may be required by normal infants with the highest requirement. Presumably, they overestimate the requirement of most infants. We have no estimates of the proportion of infants whose requirements may fall near this level or much below. Neither do we understand why different infants apparently require different amounts nor are the relationships between age and rate of growth well defined. Finally, it should be clear that these estimates are based upon work with healthy infants. What they may be in infants with various diseases or those recovering from a disease has not been directly studied.

After infancy the data are sparse indeed. Other than the limited data available from children age 10 to 12 years, the published estimates of amino acid needs are based upon extrapolation, assuming that amino acid needs change in proportion to somewhat better defined protein needs. While this seems logical, such data as are available suggest that essential amino acid requirements may fall faster than the total protein needs.

The fact that we find it necessary to publish tables of "Requirements" does nothing to improve the caliber of the data. They should be used as "best estimates" with the limitations clearly recognized.

REFERENCES

Ahrens, R.A., Wilson, J.E., Jr. and Womack, M. J. Nutr., 88, 219, 1966.

Ashworth, A. Nutr. Rev., 28, 279, 1970.

Calloway, D.H. and Margen, S. J. Nutr., 101, 205, 1971.

Fomon, S.J. and Filer, L.J., Jr. in Amino Acid Metabolism and Genetic Variation (E.L. Nyhan, ed.). McGraw-Hill Book Co., New York, 1967, p. 391.

Graham, G.G., Baertl, J.M. and Cordano, A. Amer. J. Clin., 18, 16, 1966.

Graham, G.G. in <u>Amino Acid Metabolism and Genetic Variation</u>
(W.L. Nyhan, ed.). McGraw-Hill Book Co., New York, 1967, p. 403.

Holt, L.E., Jr. and Snyderman, S.E. in <u>Amino Acid Metabolism and Genetic Variation</u> (W.L. Nyhan, ed.). McGraw-Hill Book Co., New York, 1967, p. 381.

Metta, V.C., Firth, J.A. and Johnson, B.C. <u>J. Nutr.</u>, <u>71</u>, 332, 1960.

Nakagawa, I., Takahashi, T. and Suzuki, T. <u>J. Nutr.</u>, <u>73</u>, 186, 1961.

Nakagawa, I., Takahashi, T. and Suzuki, T. <u>J. Nutr.</u>, <u>74</u>, 401, 1961.

Nakagawa, I., Takahashi, T., Suzuki, T. and Kobayashi, K.
<u>J. Nutr.</u>, <u>77</u>, 61, 1962.

Nakagawa, I., Takahashi, T., Suzuki, T. and Kobayashi, K.
<u>J. Nutr.</u>, <u>80</u>, 305, 1963.

Rose, W.C., Coon, M.J. and Lambert, G.F. <u>J. Biol. Chem.</u>, <u>210</u>, 331, 1954.

Samonds, K.W. and Hegsted, D.M. <u>Amer. J. Clin. Nutr.</u> (in press).

Srikantia, S.G. and Gopalan, C. <u>Amer. J. Clin. Nutr.</u>, <u>18</u>, 34, 1966.

INTRAVENOUS CARBOHYDRATE TOLERANCE IN INFANCY

G.W. Chance, M.B., M.R.C.P., D.C.H. (Lon.)

Neonatal Unit, The Hospital for Sick Children

Toronto, Ontario

Parenteral carbohydrate solutions are recognized as important sources of calories for neonates and infants. However, tolerance to intravenous glucose in this age group varies from that of older children. This particularly applies to very low birthweight infants since failure to recognize their limited tolerance is a potential hazard of the current widespread trend toward parenteral support in their first two weeks of life. Studies of intravenous tolerance of monosaccharides in infants have been largely restricted to the newborn period; reports on this subject in the later stages of infancy and in sick infants are very limited. In this chapter the current knowledge concerning tolerance to intravenously administered carbohydrate will be reviewed.

1. Studies of Glucose Tolerance in Infancy

The intravenous glucose tolerance test has been extensively investigated using acute glucose loads and calculating the rate of disappearance of glucose, or K_G in healthy full-term infants, those born to diabetic mothers, larger prematures and infants born small for gestational age. Interest in intravenous glucose tolerance in the neonate was particularly stimulated by recognition of the fact that the decrease of blood sugar characteristic of the first few hours of life persisted in some infants and was apparently associated with brain damage in a small minority. Contrary to what had been anticipated Baird and Farquhar in 1962 reported that despite their hypoglycemic tendency, newborns disposed of a glucose load slowly. These findings were subsequently confirmed by Bowie, Mulligan and Schwartz (1963), Cornblath, Wybregt and Baens (1963) and von Euler, Larsson and Persson (1964).

Finding higher glucose disappearance rates following a second dose of glucose immediately upon completion of the first test, Bowie et al (1963) suggested that delayed glucose disappearance was at least in part due to deficient insulin release to the initial load. In other infants these authors found that insulin given with the initial load accelerated glucose disappearance. This view was supported by the evidence of Isles, Dickson, and Farquhar (1968) who showed that rapid glucose disappearance following intravenous injections in infants of diabetic mothers was related to abnormally high insulin responses. Recently, Lerner and Porte (1971) have shown in adults that the magnitude of response of the storage pool of insulin is an important determinant of intravenous glucose tolerance. However, there remains controversy so far as infants are concerned. Thus, Gentz, Warrner, Persson and Cornblath (1969), studying intravenous glucose tolerance in appropriate and small for gestational age prematures, were unable to relate the magnitude of either acute of secondary insulin response to glucose disappearance rates in their subjects.

As a generalization, reports of intravenous glucose tolerance tests in infancy have lacked adequate standardization of conditions with regard to such aspects as the dose of glucose administered, the duration of fast, the sites of administration and sampling and the method of calculation of K_G. Despite this criticism reported studies may be summarized by saying that in the first 24 hours of life, the mean K_G following a glucose load in healthy term infants is approximately 1% per minute increasing progressively to reach adult values with a mean of approximately 2% per minute by 6 months of age. Infants of diabetic mothers generally have higher K_G values in the first 24 hours of life which have been related to the necessity for maternal insulin therapy (Isles et al, 1968), to the magnitude of their insulin response and to the degree of overweight at birth (Pedersen, 1972). Among infants who are small for gestational age and those with significant hypoglycemia have abnormally rapid K_G values, some of which can be related to increased insulin responses (Le Dune, 1972). Gentz, Persson and Zetterstrom (1969) noted that neonatal hypoglycemia resulting in neurological symptoms was associated with a high K_G value, in the majority exceeding 2 standard deviations of K_G for normal infants of the same age, whereas normal values occurred in asymptomatic infants with similar blood glucose levels.

At the Hospital for Sick Children, Toronto, we became interested in intravenous glucose tolerance in low birth weight infants during the course of a study of supplemental parenteral nutrition conducted by Drs. Heather Bryan and Patrick Wei. 10% dextrose and 10% dextrose with 3.5% fibrin hydrolysate as an intravenous supplement were compared to nasogastric formula during the first three weeks of life in infants weighing less than 1300 g at birth. Babies with severe respiratory distress syndrome or anomalies were excluded. In the first phase of the study, the total oral plus intravenous fluid intake was 150 ml per kg on the first day (day 1 - 3 of life) rising to 180 mg per kg by the third day of infusion.

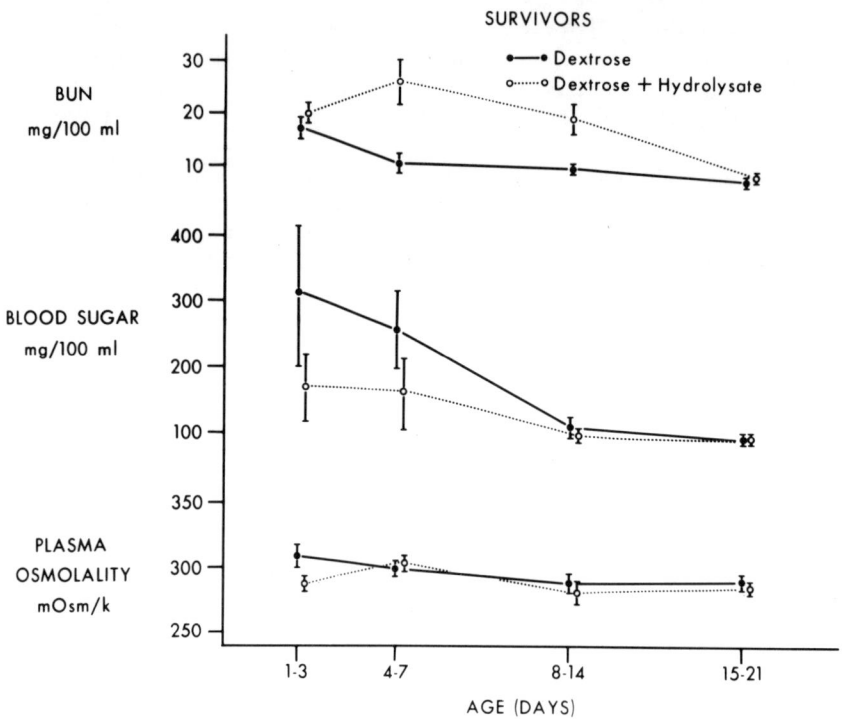

Fig. 1. Blood sugar, BUN and plasma osmolality related to age in 19 surviving infants with less than 1300 g birth weight. The initial values obtained before administration of protein supplement (mean $\pm$ SE).

The mean blood sugar levels in the 19 surviving infants of the study during their first three weeks of life are shown in Figure 1. In those newborns receiving 10% dextrose supplement, the mean was approximately 300 mg per 100 ml (ranging up to 1000 mg per 100 ml) for the first week.

The reasons for allowing such high glucose values to persist requires explanation. The blood sugar values were estimated in the routine laboratory on microsamples using the automated ferricyanide method. Unless specimens are diluted, this method becomes inaccurate at sugar values above 250 mg per 100 ml. Specimens giving this result were reported as such and not repeated on diluted samples. Also, when the studies commenced the prime attention so far as glucose in newborns was concerned was in the prevention of hypo- rather than hyperglycemia. It was only later, when checking plasma osmolalities on stored specimens from infants who died that the true extent of the hyperglycemia was recognized. The high osmolalities led us to redetermine plasma glucose using the more accurate glucose oxidase method.

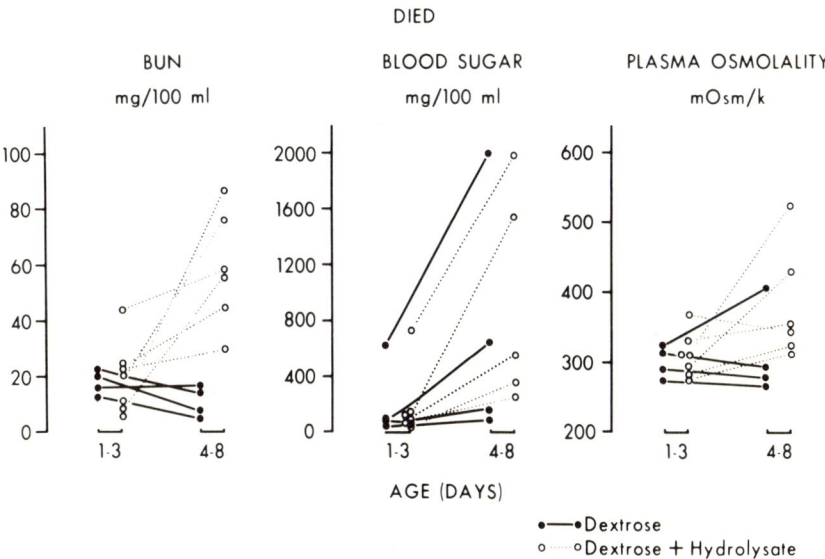

Fig. 2. Individual BUN, blood sugar and plasma osmolality during the first 8 days of life for the two types of intravenous infusion in babies who died. Day 1-3 represents pre-protein infusion babies.

The plasma osmolalities and blood sugar values in infants who died are seen in Figure 2. Very few values were in the normal range, most being above 400 mg per 100 ml. Initial values were obtained on the first day of study, and final values on the fourth to eighth day.

Further confirmation that the total sugar loads had been in excess of the infants' metabolic need and capacity to assimilate them was obtained from examination of the extent of glycosuria.

Figure 3 shows the total sugar intake and urinary sugar loss according to age. Each age group was divided according to blood sugar levels above or below 150 mg per 100 ml.

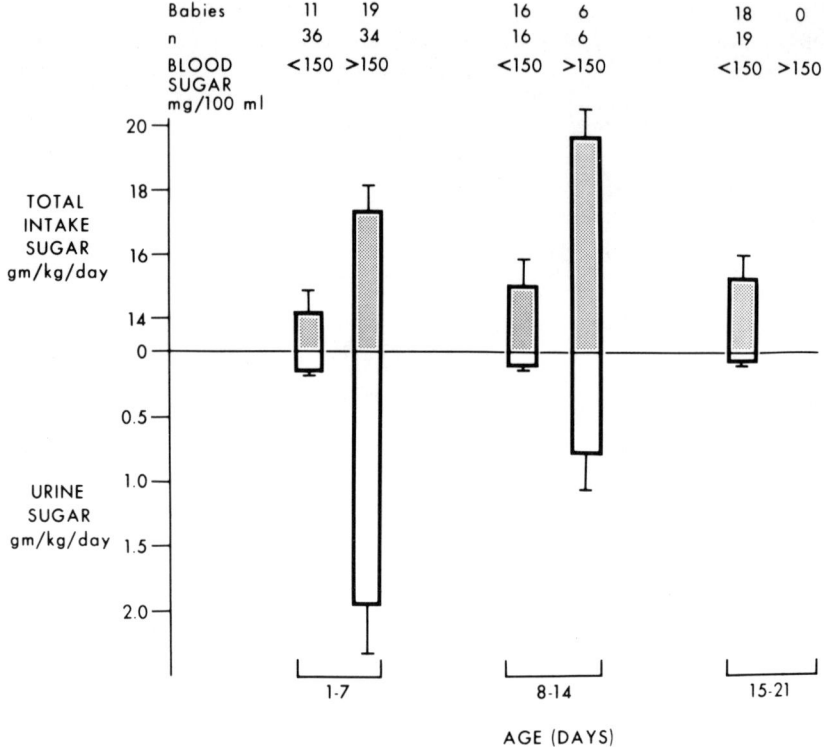

Fig. 3. Total sugar intake and uring sugar excretion (mean ± SEM) in all infants (A.G.A.) given supplemental 10% dextrose ± fibrin hydrolysate during the first three weeks of life related to blood sugar values less than or greater than 150 mg per 100 ml.

Three points I wish to stress are:
1. Glycosuria was uncommon at plasma glucose levels below 150 mg per 100 ml, and usual above this value.
2. Significant glycosuria was seen in the first week when total sugar intake (oral plus intravenous) exceeded 12 g per kg per day.
3. Sugar tolerance improved in those who survived into the second week.

Needless to say, following this experience the total fluid intake was modified, aiming to achieve a combined total oral and intravenous intake of 120 ml per kg per day for the first five days.

We studied a further 24 infants less than 1300 g similarly in all other respects monitoring blood sugar and urine sugar carefully and modifying the glucose concentration infused accordingly. We obtained much more satisfactory results.

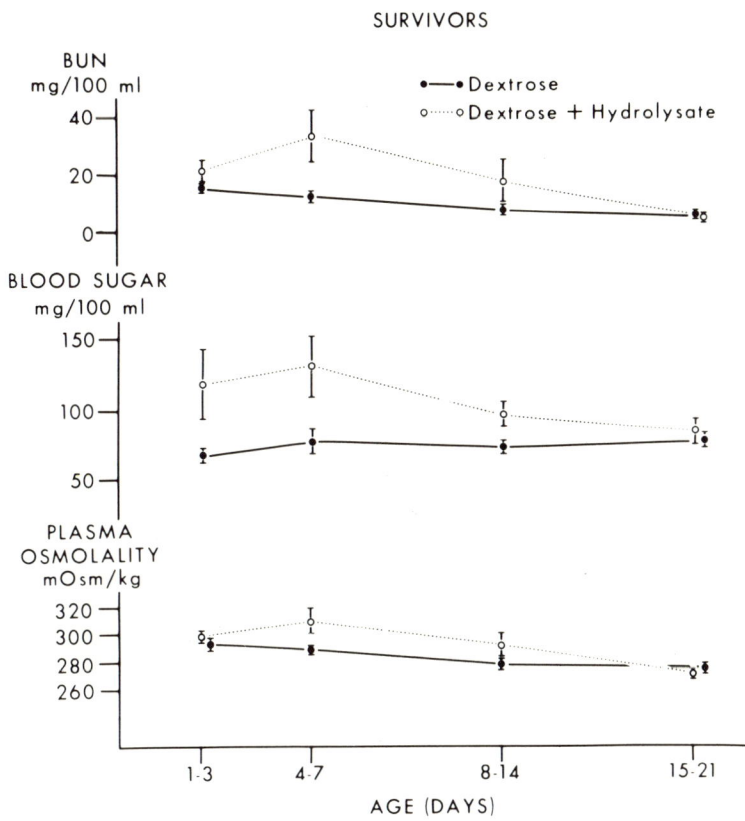

Fig. 4. Mean BUN, blood sugar and plasma osmolality related to age in surviving infants receiving a total of 120 ml per kg per day combined oral per intravenous intake.

The mean blood sugar for the survivors was maintained below 150 mg per 100 ml (Figure 4). Those receiving 10% dextrose supplements only had a mean blood sugar between 60 and 80 mg per 100 ml during the first week. In view of the work of Grasso, Messina, Saporito and Reitano (1968) showing good insulin responses to infused amino acids in prematures, it is interesting to note that the additional amino acids received by the infants we studied were not associated with improved glucose tolerance in the long-term infusion. Likewise, no hyperglycemia was encountered in any of those who died (Fig. 5).

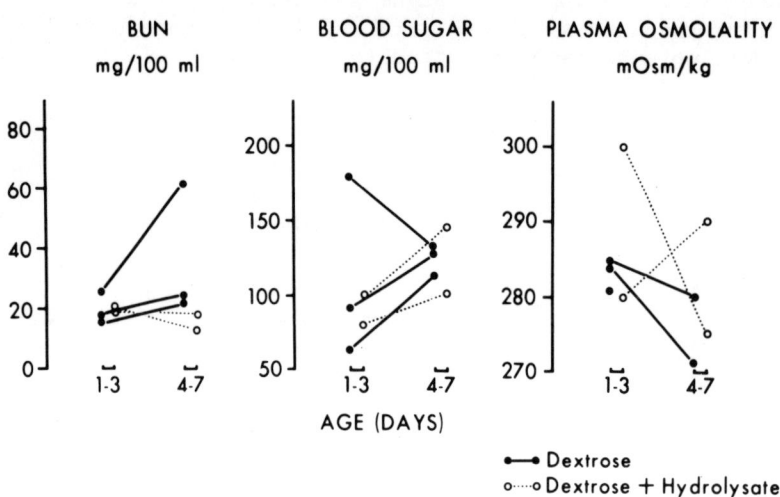

Fig. 5. Individual BUN, blood sugar and plasma osmolality in infants who died during the second phase of the study.

Stimulated by these experiences and the paucity of the literature regarding glucose tolerance in very low birth weight infants, Dr. John O'Brien and I have recently performed intravenous glucose tolerance tests and 4 hour glucose infusion studies in such infants. In performing these we have standardized the method as far as possible by fasting all infants for 4 hours, maintaining them in a thermostable 'servo' controlled environment, using a dose of 0.5 gm per kg glucose over 3 minutes, using only arterial or capillary samples as recommended by Butterfield, Abrams and Whichelow (1971) and using absolute, rather than excess glucose values obtained between 20 and 60 minutes for K_G calculations, a precaution also recommended by these authors and previously by Ikkos and Luft (1957). K_G was calculated from a semilogarithmic plot of glucose value against time as recommended by Conard (1955). Blood glucose was estimated on microsamples using a Beckman analyser.

Eighteen appropriate and 20 small for gestational age (A.G.A.

and S.G.A.) infants all weighing less than 1300 g at birth were studied, 19 on as many as four occasions up to 7 weeks of age. All were healthy at the time of study. The results of K_G plotted against $\log_{10}$ age for the A.G.A. infants are shown in Figure 6.

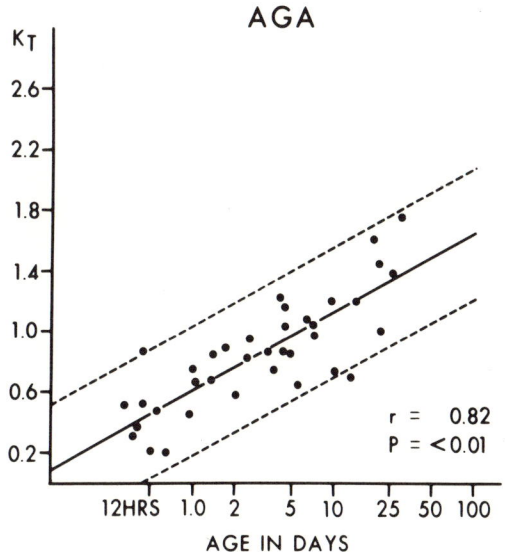

Fig. 6. K_G in 18 infants appropriate for gestational age of less than 1300 g birthweight related to $\log_{10}$ age. (r = 0.82, p is less than 0.01).

In the first 48 hours and especially the first 24 hours, K_G values were well below those quoted for more mature infants. Interestingly however, tolerance matured rapidly so that K_G comparable to those for 6 months-old infants were achieved by the age of 2 months. Contrary to the finding of Gentz et al (1969) the mean values for K_G in 7 S.G.A. infants were significantly higher than those for 11 A.G.A. infants in the first 24 hours (Table 1).

Table 1: Infants Less Than 1300 GM Birth Weight
Age Less Than 24 Hours

	K_G ($\bar{x}$)	1 SD
Infants with appropriate weights (11)	0.42	± 0.21
Infants small for gestational age (7)	0.85	± 0.31

Mean ± S.D. for S.G.A. and A.G.A. infants with less than 1300 gm birthweight in first 24 hours (P less than 0.05 and greater than 0.01).

With subsequent growth mean K_G values in S.G.A. infants did not differ significantly from those in A.G.A. infants although the scatter was somewhat greater (Figure 7).

Although statistical analysis has given a smooth and highly significant correlation with $\log_{10}$ age, I would hasten to state that this pattern is not seen in all infants during their first two weeks. Many of the problems encountered by these low weight

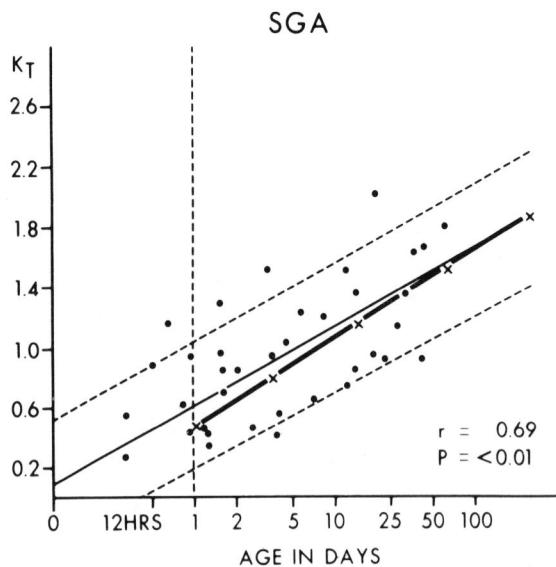

Fig. 7. K_G in 20 infants small for gestational age of less than 1300 g birthweight related to $\log_{10}$ age after first 24 hours (r = 0.69, p less than 0.01).

neonates such as those of respiratory adaptation, sepsis, etc., may profoundly disturb glucose tolerance. Dweck, Siegal and Cassady (1972) recently reported K_G values following intravenous glucose in infants who had received prior parenteral glucose than in those who had. Nearly all the infants we studied had received oral feeds and/or intravenous glucose before the 4 hour fast. The mean K_G for infants reported by Dweck et al was 1.16% per minute which is considerably higher than the values we obtained. Their values for K_G were calculated on the basis of two samples obtained between 20 and 60 minutes. Several authors have pointed out that distribution of glucose to the entire extra-cellular space may not be complete until 20 minutes after glucose loading. In many infants we studied, the 10 minute value which we obtained was off the straight line portion of the semilog plot and would have given an erroneously high K_G.

In our study the relationship between K_G and $\log_{10}$ age was not modified by gestational age, head circumference, length, weight, or skinfold thickness. The correlation between K_G and weight almost reached significance (r = 0.443, p is less than 0.05) but this was entirely due to the mutual dependence of K_G and weight on age.

To further assess variations in glucose tolerance and physiological responses to intravenous glucose in these very low birth-

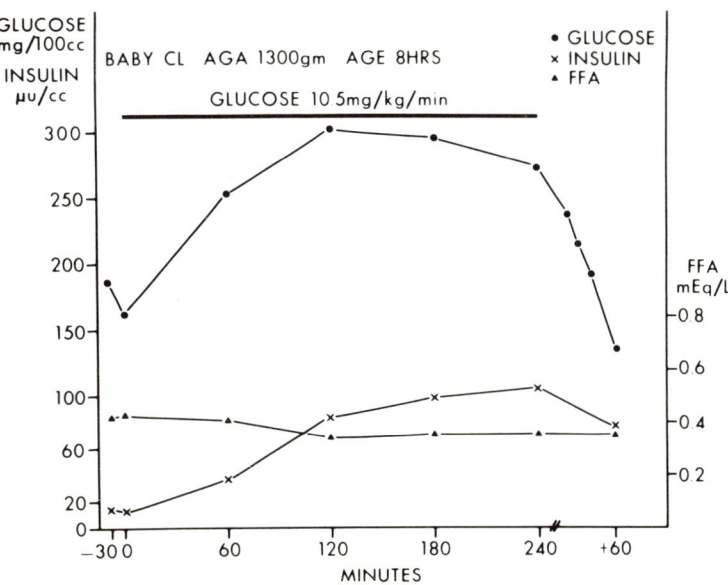

Fig. 8. Plasma glucose, insulin and free fatty acid responses to 10% dextrose solution infused at 6 ml per kg per hour in a healthy infant of 1300 g birthweight.

weight infants, Dr. O'Brien and I in collaboration with Dr. J.M. Martin recently studied responses to infusions of 10% dextrose at rates of approximately 10 mg per kg per minute over 4 hour periods.

The results have been variable but have fallen into three basic patterns as illustrated in Figures 8, 9, and 10. The first infant (Figure 8) was a healthy A.G.A. baby weighing 1300 g at birth. The initial glucose value was high, rising to 300 mg% at 2 hours. However, there was a good late insulin response and blood sugar had fallen to 270 mg per 100 ml by the end of 4 hours. Glucose disappearance was rapid when the infusion was stopped at 4 hours. Free fatty acids (F.F.A.) changed little throughout the infusion.

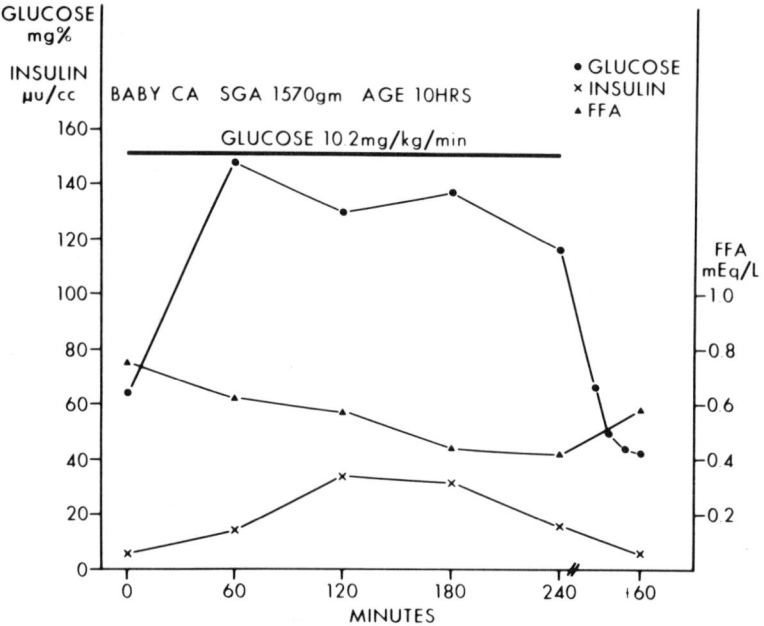

Fig. 9. Plasma glucose, insulin and free fatty acid responses to 10% dextrose solution infused at 6 ml per kg per hour in a small for gestational age infant birthweight 1570 g.

The responses obtained in an S.G.A. infant weighting 1570 g at birth are shown in Figure 9. In this infant blood glucose values rose for the first hour but fell throughout the remainder of the period. The insulin response was not marked but appears to have influenced F.F.A. release in a physiological manner. This type of response appears fairly characteristic of the S.G.A. infants we have studied to date.

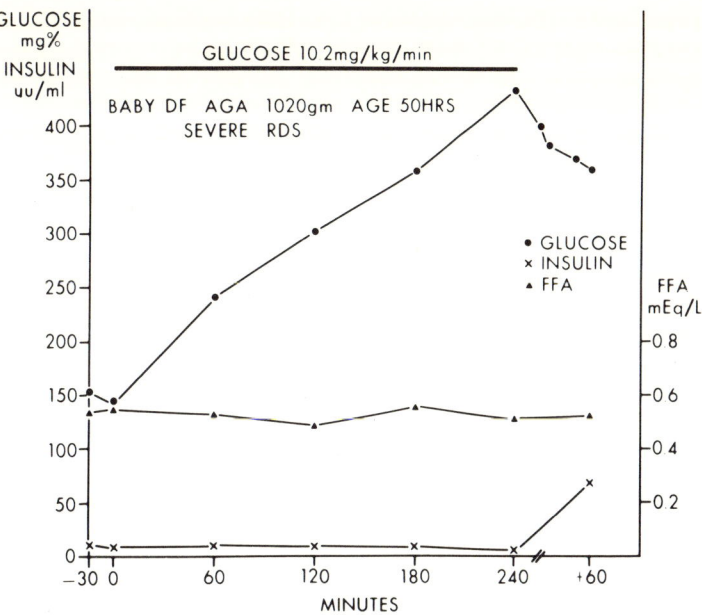

Fig. 10. Plasma glucose, insulin and free fatty acid responses to 10% dextrose infused at 6 ml per kg per hour in an A.G.A. infant with severe R.D.S. birthweight 1020 g.

The response of an infant with severe respiratory distress syndrome weighing 1020 g at birth is shown in Figure 10. There was no insulin response until the last hour of the test, glucose accumulated in the plasma throughout the study and free fatty acids were unaffected by the infusion. Unlike the last two infants, glucose assimilation was markedly reduced when the infusion was discontinued.

Despite the extensive information on K_G in normal and S.G.A. neonates and infants, there is a lack of information in regard to sick infants. Also of importance are the reports of K_G infants suffering from Kwashiorkor or marasmus. Protein malnutrition results in a reduced K_G which has been related to the poor insulin response to glucose seen in this condition (Milner, 1971, and Godard and Zahnd, 1971) whereas such changes are not a feature of marasmus. Those involved in studies of parenteral nutrition will recognize that glycosuria often signals the onset of infection but I have not been able to find systematic studies of K_G and insulin responses during infection in infants. It seems likely however, that as in the acutely burned or shocked adult, insulin release may be inhibited by release of catecholamines (Allison, Prowse, Chamberlain, 1967, Allison, Hinton, and Chamberlain, 1968).

2. <u>Intravenous Tolerance to Fructose and Galactose</u>

Both galactose and fructose disappear more rapidly than glucose from the blood following intravenous injection in healthy term infants and prematures (Mulligan and Schwartz, 1962, Cornblath, Wybregt and Baens, 1963). As with glucose however, disappearance of both of these sugars from the blood is more rapid in older children and adults than in newborns.

Schwartz and his colleagues (1964) investigated fructose tolerance in neonates up to 14 days of age and reported a progressive rise of mean K from 1.61% per minute in infants less than 6 hours of age to 3.17% per minute in infants age 7-14 days. These and other authors noted approximately a three-fold rise in plasma lactate 30 minutes following intravenous fructose loading which did not decrease with increasing age. Another feature of fructose loading reported by Schwartz et al (1964) is a pronounced early fall in blood glucose. This again is age-related being more pronounced immediately following birth. As a result of their studies, Schwartz et al suggested that this phenomenon was due to inhibition of hepatic glucose release. A French group (Odievre et al, 1970) has recently reported insulin release following a fructose load although evidence on this is conflicting. Samols and Dormandy (1963), for example, claimed that plasma insulin actually fell following fructose loading. Longer term fructose infusions appear not to be accompanied by hypoglycemia in infants and children. However, Lane and Dodd (1957) reported considerable fructosuria with a threshold as low as 10 mg per 100 ml which occurred throughout the infusion. Kaye and co-workers (1957) also found that 6 hour infusions in infants were accompanied by acidosis and hepatomegaly.

Clearance of galactose is rapid following intravenous loading in the newborn. Cornblath and colleagues (1963) reported mean values of 3.18% per minute in the first three days and 3.85% per minute up to 3 weeks. These values are very similar to those of Donnell and his colleagues (1967) and somewhat lower than those reported for adults (Colcher, Patek and Kendall, 1946). Unlike fructose, galactose infusion does not result in hypoglycemia, indeed, blood glucose levels rise rapidly in response to the infusion (Mulligan and Schwartz, 1962). Although in newborns lactate levels rise somewhat following galactose loading (Cornblath, Wybregt, Baens, 1963), this tendency does not persist in early infancy (Schwartz,

Ashmore and Renold, 1957) when in fact appearance of lactic acidosis following galactose loading is used in investigation of suspect glucose-6-phosphatase deficiency. I have been unable to find studies of long term galactose infusions in infancy but it is well tolerated in adults though urinary losses are considerable (Felber, Renold and Zahnd, 1959).

Summary

The literature on intravenous glucose tolerance tests in full term and premature newborn infants has been reviewed. Unlike amino acids, glucose fails to induce an acute insulin response with resultant glucose intolerance and low K_G values. In very low birth weight babies this early intolerance may be extreme and is reflected in an inability to tolerate high infusion glucose rates when glucose is used as a parenteral calorie source. If it goes unrecognized this intolerance may result in dangerous hyperglycemia. Results of acute intravenous loading tests with fructose and galactose show that these sugars are cleared more rapidly than glucose by the neonate. Since both result in other undesirable biochemical changes their place in boosting the neonate's parenteral calorie intake requires careful study before they can be recommended for widespread use.

Acknowledgements

The author wishes to acknowledge the technical assistance of Miss E. Denoga.

This work was supported by grants from the A. S. Hellyer Foundation and the Paediatric Consultants Staff Fund of the Hospital for Sick Children, Toronto, Canada; both of which are gratefully acknowledged.

REFERENCES

Allison, S. P., Hinton, P., Chamberlain, M. J. Lancet, 2:1113, 1968.

Allison, S. P., Prowse, K., Chamberlain, M. J. Lancet, 1:478, 1967.

Baird, J. D. and Farquhar, J. W. Lancet, 1:71, 1962.

Bowie, M. D., Mulligan, P. B. and Schwartz, R. Pediatrics, 31:590, 1963.

Butterfield, W. J. H., Abrams, M. E. and Whichelow, M. J. Metabolism, 20:225, 1971.

Colcher, H., Patek, A. J. and Kendall, F. E. J. Clin. Invest., 25 768, 1946.

Conard, V. Acta gastro-Enterol. Belg., 18:689, 1955.

Cornblath, M., Wybregt, S. H. and Baens, G. S. Pediatrics, 32: 1007, 1963.

Donnell, G. M., Ng, W. G., Hodgman, J. and Bergren, W. R. Pediatrics, 39:829, 1967.

Dweck, H. S., Siegal, A. M. and Cassady, G. Pediat. Res., 6:409, 1972.

Felber, J. P., Renold, A. E. and Zahnd, G. R. Mod. Probl. Paediat., 4:467, 1959.

Gentz, J., Persson, B. and Zetterstrom, R. Acta Paediat. Scand., 58:481, 1969.

Gentz, J., Warrner, R., Persson, B. E. M. and Cornblath, M. Acta Paediat. Scand, 58:481, 1969.

Godard, C. and Zahnd, G. R. Helv. Paediat. Acta, 26:276, 1971.

Grasso, S., Messina, A., Saporito, N. and Reitano, G. Lancet, 2: 755, 1968.

Ikkos, D. and Luft, R. Acta Endocrinol., 25:312, 1957.

Isles, T. E., Dickson, M. and Farquhar, J. W. Pediat, Res., 2:198, 1968.

Kaye, R., Williams, M. L. and Barbero, G. Am. J. Dis. Childh., 93: 85, 1957.

Lane, H. C. and Dodd, K. Pediatrics, 20:668, 1957.

Le Dune, Marthe A. Arch Dis Childh., 47:111, 1972.

Lerner, R. L. and Porte, D. J. Clin. Endocrinol., 33:409, 1971.

Milner, R. D. G. Pediat. Res., 5:33, 1971.

Mulligan, Paula B. and Schwartz, P. Pediatrics, 30:125, 1962.

Odievre, M., Poirier, C., Levillain, P., Modigliani, E., Strauch, G. Arch. Franc. Ped., 27:1057, 1970.

Pedersen, L. M. Acta Endocrinol., 69:174, 1972.

Samols, E. and Dormandy, T. L. Lancet, 1:478, 1963.

Schwartz, R., Ashmore, J. and Renold, A. E. Pediatrics, 19:585, 1957.

Schwartz, R., Gamsu, H., Mulligan, Paula B., Reisner, S. H., Wybregt, Susan, H. and Cornblath, M. J. Clin. Invest., 43:333, 1964.

Von Euler, U., Larsson, Y. and Persson, B. Arch. Dis. Childh., 39:393, 1964.

SUBSTRATE SUPPLY AND UTILIZATION IN VARIOUS CONDITIONS

Julien F. Biebuyck

Department of Anesthesia

Harvard Medical School, at the Massachusetts General Hospital and Shriners Burns Institute, Boston, Massachusetts

The main substances which can be directly utilized as a source of energy in mammalian tissues are (Krebs, 1972):

(1) Glucose (fructose, galactose).
(2) Fatty acids, including acetate.
(3) Ketone bodies derived from fatty acids.
(4) Amino acids.
(5) Lactate derived from glucose by glycolysis.

Not every organ can utilize every one of these substances. The brain and blood cells depend on glucose as the main fuel. The liver, kidney cortex, and both cardiac and skeletal muscle obtain most of their energy from the oxidation of free fatty acids. In addition, ketone bodies can make a major contribution to the fuel of respiration in brain, renal cortex, and cardiac skeletal muscle.

REGULATING FACTORS IN FUEL UTILIZATION BY TISSUES

The main factors (Krebs, 1972) affecting which source of energy is used are as follows:

(a) the release of substrates from tissue stores of triglycerides and glycogen
(b) the concentration of the fuel in blood plasma
(c) the entry of the fuel from the plasma into tissues
(d) the presence in the tissue of enzymes required for the degradation

These factors are influenced by hormones such as insulin, epinephrine, glucagon, and corticosteroids, and by cyclic-AMP.

CARBOHYDRATE AS A DIRECT FUEL

Although carbohydrate is often the main ultimate source of energy in the organism as a whole, the quantitative role of glucose as a direct source of energy is a relatively minor one. This is due to the fact that the amount of fuel which can be stored as carbohydrate is very limited. The energy that can be stored as glycogen in liver and muscle is rather less than one day's calorie need (Cahill, 1970). During fasting, fatty acids are released from the adipose tissue to serve as a fuel.

ADVERSE EFFECTS OF THE INTRAVENOUS USE OF FRUCTOSE

The advantages of fructose over glucose have led to its use as a fuel in intravenous feeding. Fructose is more rapidly metabolized than glucose and does not depend on insulin for intracellular penetration (Miller et al., 1952). Fructose is less irritating to the peripheral veins and can be given in higher concentrations than glucose (Thoren, 1964). Uremic and postoperative patients with glucose intolerance appear to be able to utilize fructose normally (Luke et al., 1964). For these reasons, fructose has been used in patients with uremic or diabetic ketoacidosis and post-operatively.

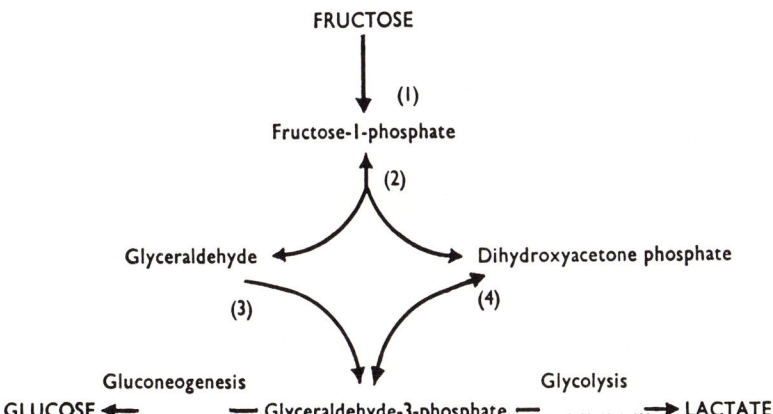

Fig. 1 Main pathway of fructose metabolism in liver. The enzymes involved in these reaction steps are: (1) fructokinase, (2) ketose-1-phosphate aldolase, (3) triokinase, and (4) triosephosphate isomerase. (From Woods and Alberti, 1972)

However, it has become apparent (Bergström et al., 1971) that in healthy subjects fructose infusion raises the blood lactate concentration (see Figure 1).

This rise in blood lactate after infusions of fructose in low concentrations (0.5g. per Kg. body weight per hour) is large enough to disturb the acid-base balance in clinical settings associated with acid-base disturbances. Fructose is rapidly phosphorylated in the liver and leads to a sharp fall in liver ATP concentrations (Woods and Alberti, 1972; Maenpaa et al., 1968). It has been suggested that ATP is utilized for fructose phosphorylation in preference to protein synthesis, and that this results in an interference with uric acid metabolism.

INTERRELATIONSHIPS OF CARBOHYDRATE AND FAT METABOLISM: "CALORIC HOMEOSTASIS"

The term "caloric homeostasis" was coined by Fredickson and Gordon (1958) to express the joint and complementary roles which free fatty acids and glucose play in the blood in supplying the respiration fuel. The interconversion of carbohydrate and fat has been described by Randle et al. (1963) as the glucose-fatty acid cycle (see Figure 2) Reactions 1 and 2 occurs when there is a surplus of carbohydrate (or of ingested fat). Reaction 3 occurs on fasting. The plasma non-esterified fatty acid

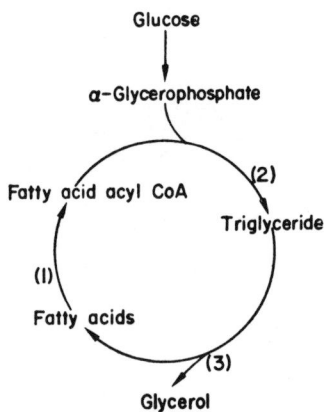

Fig. 2 The "glucose fatty acid cycle" according to Randle et al., 1963.

concentration rises whenever the blood glucose is low, as in starvation or on a low-carbohydrate diet, or when glucose is not readily available to the tissues, as in diabetics. The increase in free fatty acids in these situations is accompanied by a parallel increase in the concentrations of ketone bodies.

FAT AS A FUEL

In the post-absorptive state stored fat is presented to the tissues in four forms: free fatty acids derived directly from adipose tissue; triglycerides formed in the liver from fatty acids; and acetoacetate; and hydroxybutyrate formed by the partial oxidation of free fatty acids. Free fatty acids and ketone bodies are readily oxidizable. It may be that more than one form is necessary because free fatty acids alone, owing to their low solubility and their relative toxicity, cannot be transported in sufficient quantities to meet the fuel requirements of some tissues, such as heart muscle. The other possibility for the existance of multiple forms of this substrate is that it allows direction of particular fuels to specific organs.

Each of the above requires a different enzyme to initiate its utilization: acyl-CoA synthetase for free fatty acids; lipoprotein lipase for triglycerides; 3-hydroxybutyrate dehydrogenase for hydroxybutyrate. The potential of any tissue to metabolize these fuels will therefore depend on its content of the necessary enzymes, and this may vary with the physiological state of experimental animals.

THE ROLE OF KETONE BODIES

It is well established that many animal tissues can oxidize ketone bodies (Snapper and Grunbaum, 1927; Wick and Drury, 1941; Williamson and Krebs, 1961), and this has led to the concept that it is a physiological function of ketone bodies to serve as a fuel of respiration when carbohydrate is in short supply (Krebs, 1961). Experiments have shown that increased production of ketone bodies is closely matched by increased utilization (Bates et al., 1968). These authors suggest the sequence of events leading to "physiological ketosis": as a consequence of hormonal interrelationships a low blood sugar concentration causes an increase in adipose tissue lipolysis and a rise in the concentration of free fatty acids in the plasma. This in turn results in an increased rate of ketogenesis in the liver, which is followed by a rise in blood ketone-body concentrations, and an increased rate of peripheral utilization.

The formation of ketone bodies (D-hydroxybutyrate and acetoacetate) results from the partial oxidation of fatty acids. (see

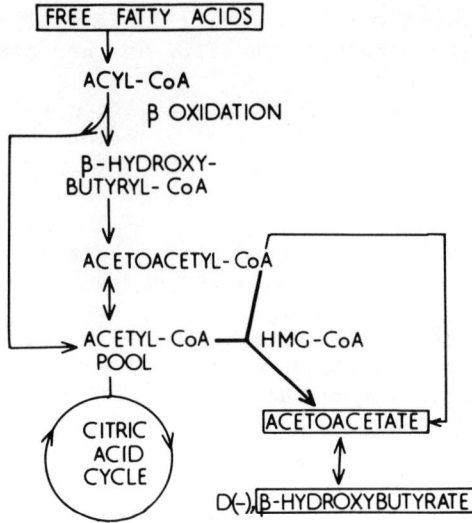

Fig. 3 Diagram to show pathways of ketogenesis in the liver. The formation of ketone bodies (acetoacetate and 3-hydroxybutyrate) in the liver, by partial oxidation of free fatty acids.

Ketogenesis required three specific reactions which have nothing directly to do with fatty acid degradation (Krebs et al., 1971). These reactions are the HMG-CoA synthase reaction, the HMG-CoA lyase reaction, and the 3-hydroxybutyrate dehydrogenase reaction:

(1) Acetoacetyl-CoA + acetyl-Co A → hydroxymethylgutaryl-CoA + CoA
(HMG-CoA synthase)

(2) Hydroxymethyglutaryl-CoA → acetoacetate + acetyl-CoA
(HMG-CoA lyase)

(3) Acetoacetate + $NADH_2$ ⇌ 3-hydroxybutyrate + NAD
(3-hydroxybutyrate dehydrogenase)

The significance of hepatic ketogenesis lies in the supply of a fuel of respiration to peripheral tissues (Krebs, 1961).

Ketone body utilization depends on the presence of key enzymes in the tissues:

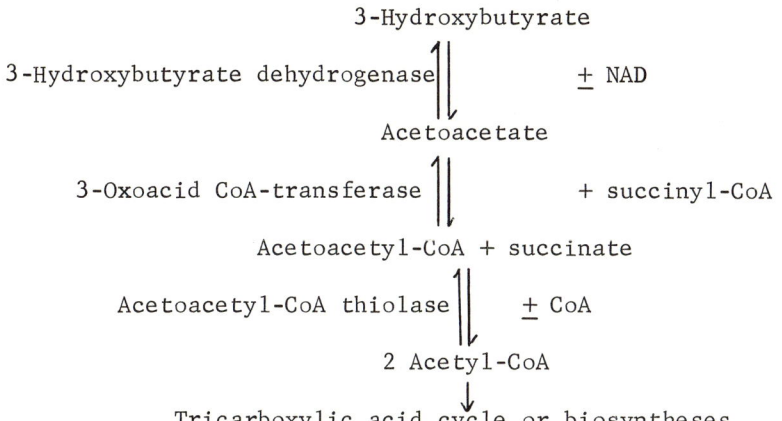

The increased rate of ketone body utilization in starvation is essentially not determined by a raised capacity of the enzymes of ketone body utilization. By far the most important factor would appear to be the concentration of ketone bodies in the blood. (See Figure 4)

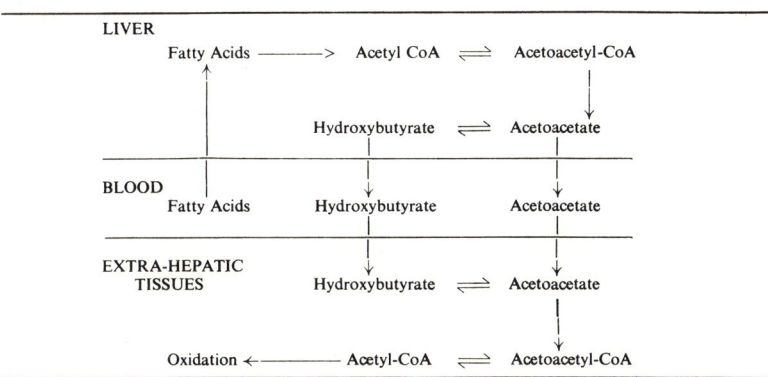

Fig. 4 Outline of ketone body metabolism. The liver is the only organ which contributes significant amounts of ketone bodies to the circulating blood. The two main ketone bodies (acetoacetate and 3-hydroxybutyrate) can be interconverted in liver and the diffuse into the blood to supply extrahepatic tissues. The liver itself is unable to utilize ketone bodies to any appreciable extent. (From Williamson and Hems, 1970).

The concept of "physiological ketosis", the sensitivity of the concentration of ketone bodies in blood to changes in the hormonal and nutritional state (particularly in the neonatal period), and the existence of rapid and specific methods for the estimation of ketone bodies (Williamson et al., 1962) suggest that a request for determination of blood ketone bodies should become a common part of our clinical practice.

REGULATORY FACTORS IN KETOGENESIS AND UTILIZATION

The regulation of ketogenesis is related to the regulation of the tissue concentration of free fatty acids, the main ketogenic precursors (Krebs and Hems, 1970; Mayes, 1970). This concentration, in turn, depends mainly of the rate at which fatty acids reach the liver from the triglyceride stores of adipose tissue and from the intestine. Several hormones--- glucagon, epinephrine, norephinehrine, and corticosteroids--- can promote the release of free fatty acids from adipose tissue. Ketone bodies may also, themselves, regulate the release of free fatty acids from adipose tissue. The rate of ketogenesis by the liver from fatty acids has been shown to increase with the substrate concentration (Heimberg et al., 1969; Krebs et al., 1969). The increased ketogenesis in starvation can in part be explained by the higher concentration of free fatty acids in the blood plasma and presumably in the tissue. However, addition of free fatty acids to the liver of well-fed rats does not raise the rate of ketogenesis to the same value as in the liver of starved rats. Another factor which may effect the rate of ketogenesis in this situation is the availability of non-fat substrates (Krebs et al., 1969).

The liver is clearly well equipped to utilize free fatty acids and to interconvert acetoacetate and hydroxybutyrate, but the virtual absence of 3-Oxoacid-CoA transferase and lipoprotein lipase means that any significant uptake of ketone bodies and triglycerides is restricted to extra-hepatic tissues. Heart and kidney contain the necessary enzymes to deal with all four fuels and this may reflect their high metabolic activity. Page and Williamson (1971) have shown that normal human brain has the capacity to utilize ketone bodies.

EFFECT OF ADMINSTRATION OF KETONE BODIES

The administration of aceoacetate or hydroxybutyrate to fasting man or animals produces a prompt decrease in the concentrations of free fatty acids in the blood (Mebane and Madison, 1964; Balasse et al., 1967; Senior and Loridan, 1968; For review, see Williamson and Hems, 1971). Two mechanisms have been proposed for the decreased flux of free fatty acids from adipose tissue on administration of ketone bodies:

(a) direct inhibition of lipolysis bought about by acetoacetate or hydroxybutyrate (Bjornthorp, 1966);

(b) stimulation of insulin secretion by ketone bodies with resultant suppression of the release of free fatty acids (Madison et al., 1964).

The decrease of free fatty acid concentration after administration of ketone bodies is accompanied by a fall in the concentrations of glucose and glycerol in the blood (Balasse and Ooms, 1968). Further work is required before the relative importance of the modification of the supply of metabolic fuels by ketone bodies can be correctly assessed.

GLUCONEOGENESIS AND KETOGENESIS IN NEONATAL DEVELOPMENT

Glycolysis, the hexosemonophosphate shunt, and lipogenesis are prominent in fetal liver, whereas gluconeogenesis is essentially absent (Adam, 1971). The pattern is essentially reflected in the levels of specific enzymes involved in these pathways (Dawkins, 1966; Burch et al., 1963). The enzymes of amino acid catabolism and of the urea cycle are also low in the fetal liver.

Immediately following birth, dynamic changes occur in the blood and tissue levels of many substances and in the rates of pathways and levels of enzymes in the liver (Figure 5).

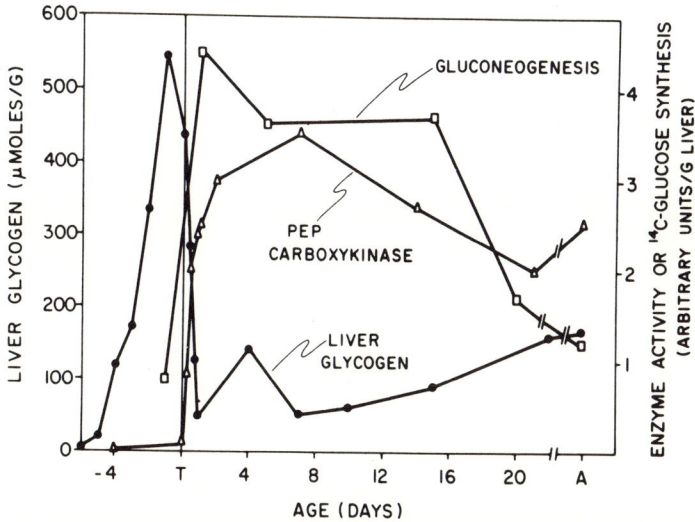

Fig. 5 Post-natal changes in the livers of rats. (From Exton, 1972)

The initial decline in the blood glucose in the neonate is explained by the very high rate of glucose utilization and, of course, by the cessation of glucose inflow from the mother.

Stabilization is achieved as a result of:

(1) increased output of glucose from the liver due initially to glycogenolysis and later gluconeogenesis

(2) replacement of glucose by FFA as a fuel for muscle

(3) resumption of carbohydrate intake

The mobilization of FFA (See Figure 6) (Van Duyne and Havel, 1959) following birth has been shown to be due to the initial hypoglycemia which suggests it probably results from activation of the sympathetic system. The change from glucose to fat as an energy source is accompanied by a fall in the RQ from approximately 1.0 in the fetus to 0.9 during the first few hours of life, and to 0.73 at the third day.

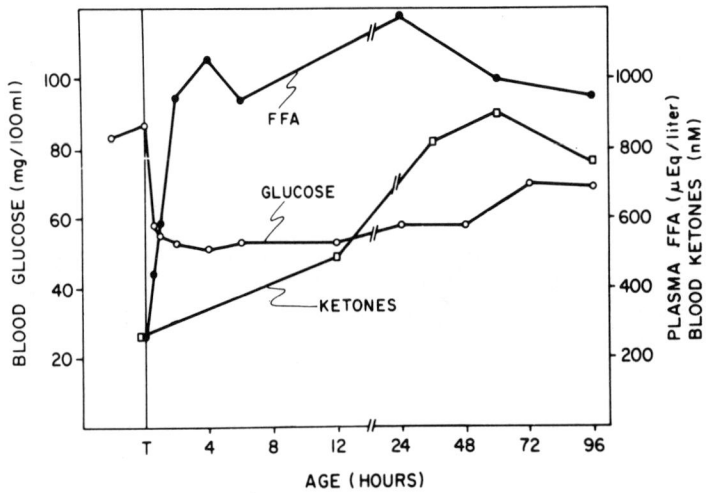

Fig. 6 Post-natal changes in blood of human infants.
(From Exton, 1972)

The increased release of FFA following birth leads to increased ketone body production by the liver. It is likely that the availability of ketones and lactate for utilization by the brain accounts for the well-known resistance of new born infants to hypoglycemia. Hepatic gluconeogenesis (See Figure 7) is negligible in the fetus and increases very rapidly during the first week of life paralleling the induction of P-Enolpyruvate Carboxykinase (Ballard and Oliver, 1963; Ballard, 1971).

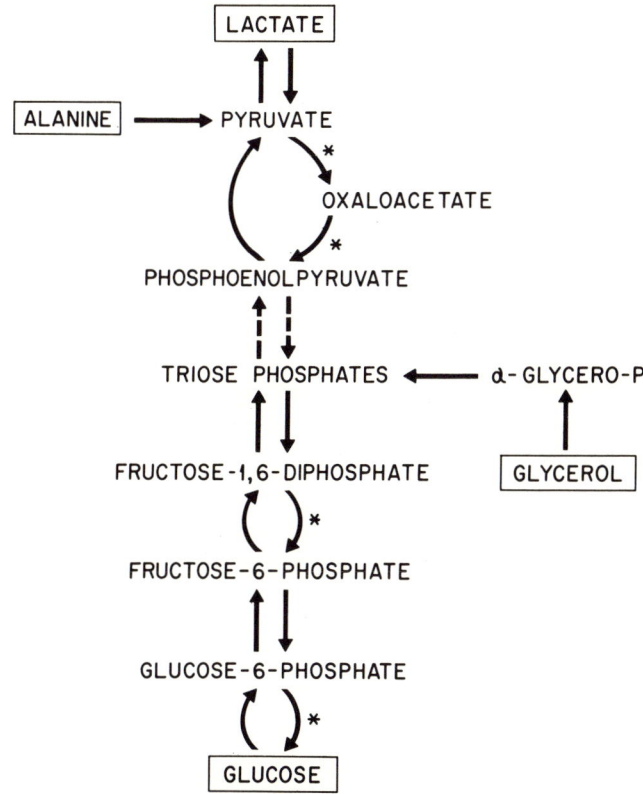

Fig. 7 <u>Diagram of the pathway of glyconeogenesis from various precursors</u>. The diagram shows the stages where amino acids (alanine), glycerol, and lactate, join the pathway of gluconeogenesis. The reactions which are common to gluconeogenesis and glycolysis are those indicated by the straight arrows, the downward arrows showing the direction of gluconeogenesis, the upward arrows the direction of glycolysis. The curved arrows represent the reactions (pyruvate carboxylase, phosphoenolpyruvate carboxykinase, fructose-1,6-diphosphatase, glucose-6-phosphatase) which circumvent the energy barriers obstructing the direct reversal of glycolysis.

The blood lactate level in the fetus is high because of relative hypoxia and because glucose is the principle fuel. Following birth, blood lactate decreases rapidly due to the initiation of gluconeogenesis, the increase in oxygenation and the decrease in peripheral glucose utilization (Exton, 1972). Lactate later becomes supplanted as a gluconeogenic substrate by amino acids and glycerol. The utilization of amino acids for gluconeogenesis is negligible prior to birth, then rapidly increases (in rat liver) reaching a peak at the 5th day, declining to the adult rate about the 30th day (Yeung and Oliver, 1967).

KETONE BODY METABOLISM IN NEONATAL PERIOD

Itoh and Quastel (1970) have shown that slices of brain of infant rats utilize ketone bodies more rapidly than slices from adult brain. It is already known (Klee and Sokoloff, 1967) that 3-hydroxybutyrate dehydrogenase activity is higher in the brain of infant rats than of adults, and declines after weaning. Striking changes in the blood ketone body concentration occur during the suckling period (Krebs et al., 1971) in young rats. At birth the concentrations were similar to those found in fed adult rats, but within one day the ketone body concentration rose to levels found in 48 hour starved adult rats, and this high concentration was maintained until the time of weaning when it decreased toward the adult fed value. The concentration of free fatty acids and of glycerol are also higher in the suckling rat than in fed adults. After weaning, i.e. after 21 days, they begin to decrease towards the adult fed values. Thus, unlike the hyperketonemia of starvation the hyperketonemia of suckling rats is accompanied by a normal, or even slightly raised, glucose concentration.

Findings in neonatal rats suggest that the development of organ enzyme patterns in the neonate is co-ordinated so as to direct ketone bodies to the brain. The circulating concentration of ketone bodies is the major factor in the control of their uptake by the brain (Hawkins et al., 1971). In suckling rats, the arterio-venous differences of ketone bodies across the brain are also proportional to the ketone body concentration, but at any given concentration the A-V differences are about 3-fold greater than in adult rats. This finding is in agreement with the raised activity of the enzymes of ketone body utilization in brain of suckling rats (Krebs et al., 1971).

ALTERATIONS IN FUEL METABOLISM DURING AND FOLLOWING SURGERY

Following trauma, accelerated nitrogen catabolism occurs and an inadequate sparing of nitrogen reserves results (Cuthbertson,

1930; Cuthbertson and Tilstone, 1968; Johnston, 1972). Alanine, and to a lesser extent glutamine, are utilized by the liver for gluconeogenesis (See Fig. 8). Trauma thus places an extra

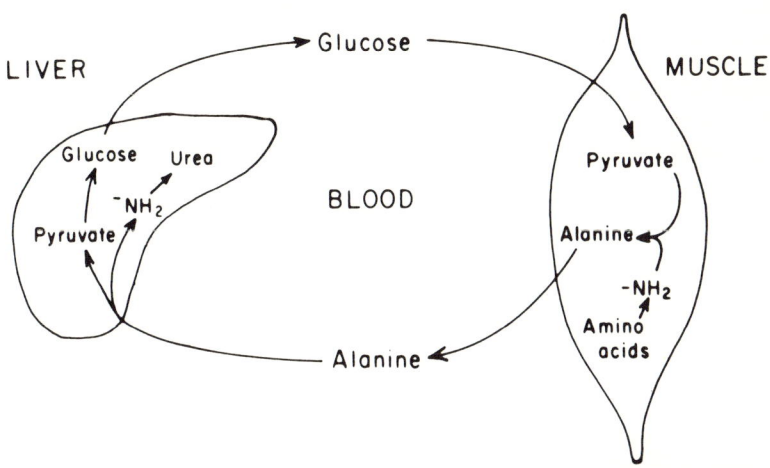

Fig. 8 Alanine cycle. Alanine serves to convey amino groups and carbon substrate from muscle to liver for conversion to urea and glucose, respectively. The alanine is re-synthesized in muscle by transamination of pyruvate derived from glucose or other amino acids. (From Exton, 1972)

gluconeogenic load on the body, apparently to provide glucose needed to reparative tissues. Administration of small or even large amounts of glucose to traumatized man usually fails to suppress this nitrogen mobilization (Cahill and Aoki, 1972). The capacity of these individuals to handle exogenous glucose is markedly decreased, and extremely severe hyperglycemia may result from over-enthusiastic exogenous glucose administration in an attempt to provide "calories."

It has been shown that during anesthesia and post-operatively, following severe myocardial infarction or burns, high levels of blood glucose are associated with less rise in plasma insulin then occurs in non-stress states (Allison et al., 1968; Allison et al., 1969).

The mechanism of the deterioration of glucose tolerance during and after operation is variable and complex, but in general, is due to factors suppressing insulin release from beta cells, and/or factors reducing the effect of insulin on the liver, muscle or adipose tissue (Soeldner, 1972; Lindseth, 1972). In contrast to these findings, Carlson (1972), has found that intravenous "fat tolerance" is considerably increased by trauma such as surgery, an opposite effect from that on glucose tolerance.

Altered glucose metabolism during anesthesia has been found by several investigators. We have shown in our laboratory (Biebuyck et al., 1972a), that anesthetic agents inhibit gluconeogenesis. Fatty acid metabolism, however, does not appear to be effected by anesthetic agents (Biebuyck et al., 1972b). We have found that, the addition of oleate to the perfusion medium of the isolated liver leads to a decrease in the inhibition by halothane (a commonly used agent in clinical anesthesia) of oxygen comsumption, urea synthesis and gluconeogenesis, and also maintains the liver content of ATP. The inhibited rate of urea synthesis caused by this drug (Seen in Figure 9) is restored to the normal range by the presence of 2mM oleate in the perfusate.

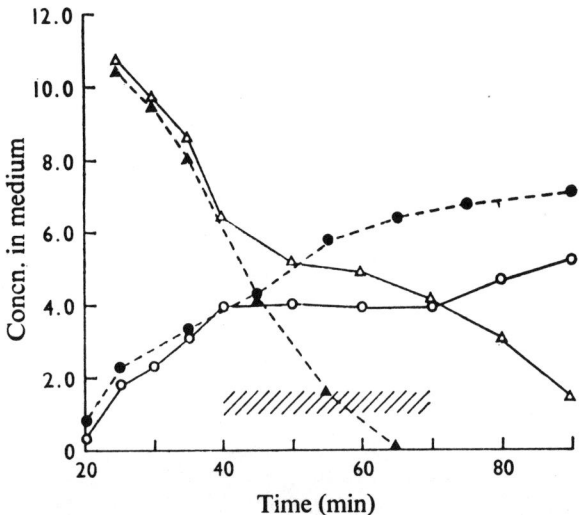

Fig. 9. Times course of urea synthesis from 10mM-NH_4Cl and 2mM-ornithine: effect of 2.5% (v/v) halothane. Livers from rats starved for 48 hours were perfused, substrates were added to the perfusion medium at 20 minutes, and samples were taken at 5 minute intervals (not shown on the figure). The hatched area indicated the period of exposure to halothane.
---Control. —— Halothane-treated.
▲ △ Ammonia removal.   ● o Urea synthesis.
(From Biebuyck, Lund, and Krebs, 1972a)

A similar lack of adverse effects of halothane has been found in liver tissue metabolic intermediates in animals fed a high fat diet, or animals starved for 24 hours (Biebuyck and Lund, 1972). Thus, situations in which a high concentration of fatty acids is available to the liver (starvation or exogenous administration), appear to have a protective effect against metabolic changes bought about by halothane.

In view of these findings it is felt that careful consideration should be given to further investigation of the use of fat emulsions or ketone body containing fluids in the neonatal period during prolonged surgical procedures, and post-operatively. Apart from the relative lack of interference with fat metabolism compared to carbohydrate metabolism during and following anesthesia, fatty acids or their products (ketone bodies) are the primary fuel source for liver, myocardium and muscle tissue.

Conclusion: These many observations can be fitted together in terms of the concept of "caloric homeostasis". Since dissimable substrate must be available for the production of energy at all times, an organ must, if it is to survive, have a means of mobilizing energy reserves when exogenous foodstuffs are in short supply. It is toward the maintenance of this homeostatic process that we should direct our attentions in planning our approach to intravenous feeding in abnormal clinical situations.

REFERENCES

Adam, P.A.J. (1971), Advances Metab. Dis.: 5, 184.

Allison, S.P., Hinton, P., and Chamberlain, M.J. (1968), Lancet: 2, 1113.

Allison, S.P., Tomlin, P.J., and Chamberlain, M.J. (1969), Brit. J. Anesth.: 41, 588.

Balasse, E., Couturier, E., and Franckson, J.R.M. (1967), Diabetologia: 3, 488.

Balasse, E. and Ooms, H.A. (1968), Diabetologia: 4, 133.

Ballard, F.J. (1971), Biochem. J.: 124, 265.

Ballard, F.J. and Oliver, I.T. (1963), Biochem. Biophys. Acta.: 71, 578.

Bates, M.W., Krebs, H.A., and Williamson, D.H. (1968), Biochem. J.: 110, 655.

Bergstrom, J., Hultman, E., and Rochnorland, A.E. (1968) Acta Med. Scand. 184, 359.

Biebuyck, J.F. and Lund, P. (1972) Annual Meeting American Society of Anesthesiologists p. 145.

Biebuyck, J.F., Lund, P., and Krebs, H.A. (1972a) Biochem. J.: 128, 711.

Biebuyck, J.F., Lund, P., and Krebs, H.A. (1972b) Biochem. J.: 128, 721.

Bjorntorp, P. (1966) J. Lipid. Res.: 7, 621.

Burch, H.B., Lowry, O.H., Kuhlman, A.M., Skerjance, J., Diamant, E.J., Lowry, S.R., and Von Dippe, P. (1963) J. Biol. Chem.: 238, 2267.

Cahill, G.F., Jr. (1970) New Eng. J. Med.: 282, 668

Cahill, G.F., Jr. and Aoki, T.T. (1972) In: Intravenous hyperalimentation (Cowan, G.S.M. and Scheetz, W.L., eds.) Philadelphia: Lea and Febiger.

Carlson, L.A. (1972) In: Intravenous Hyperalimentation (Cowan, G.S.M. and Scheetz, W.L., eds) Philadelphia: Lea and Febiger.

Cuthertson, D.P. (1930) Biochem. J.: 24, 1244.

Cuthertson, D.P. and Tilstone, W.J. (1968) Am. J. Clin. Nutri.: 21, 911.

Dawkins, M.J.R. (1966) Brit. Med. Bull.: 22, 27.

Exton, J.H. (1972) Metabolism: 21, 945.

Fredrickson, D.S. and Gordon, R.S. (1958) Physiol. Rev.: 38, 585.

Hawkins, R.A., Williamson, D.H. and Krebs, H.A. (1971) Biochem. J.: 122, 13.

Heimberg, M., Weinstein, I., and Kohout, M. (1969) J. Biol. Chem.: 244, 5131.

Itoh, T. and Quastel, J.H. (1970) Biochem. J.: 116, 641.

Johnston, I.D.A. (1972) Advances in clinical chemistry, Vol. 15 p. 255. (Bodansky, O. and Latner, A.L., eds.) New York: Academic Press.

Klee, C.B. and Sokoloff, L. (1967) J. Biol. Chem.: 242, 3880

Krebs, H.A. (1961) Biochem. J.: 80, 225.

Krebs, H.A. (1972) Advances in Enzyme Regulation, Vol. 10, p. 397, (Weber, G., ed.) Oxford: Pergamon Press.

Krebs, H.A. and Hems, R. (1970) Biochem. J.: 119, 525

Krebs, H.A., Wallace, P.G., Hems, R., and Freedland, R.A. (1969) Biochem. J.: 112, 595.

Krebs, H.A., Williamson, D.H., Bates, M.W., Page, M.A. and Hawkins, R.A. (1971) Advances in Enzyme Regulation Vol. 9, p. 387, (Weber, G., ed.) Oxford: Pergamon Press.

Luke, R.G., Dinwoodie, A.J., Linton, A.L., and Kennedy, A.C. (1964), J. Lab. Clin. Med.: 64, 731.

Lindseth, R.E. (1972) Arch. Surg.: 105, 741.

Maenpaa, P.A., Raivo, K.O., and Kekomaki, M.P. (1968) Science: 161, 1253.

Madison, L.L., Mebane, D. Unger, R.H. and Lochner, A. (1964) J. Clin. Invest.: 43, 408.

Mayes, P.A. (1970) In: Adipose tissue regulation and metabolic functions, p. 186 (Jeanrenaud, B. and Hepp, D., eds.) New York: Academic Press.

Mebane, D. and Madison, L.L. (1964) J. Lab. Clin. Med.: 63, 177.

Miller, M., Drucker, W.R., Owens, J.E., Craig, J.W., and Woodward, H., Jr. (1952) J. Clin. Invest.: 31, 115.

Page, M.A. and Williamson, D.H. (1971) Lancet:11, 66.

Randle, P.J., Garland, P.S., Hales, C.N., and Newsholme, E.A. (1963) Lancet: 1, 785.

Senior, B. and Loridan, L. (1968) Nature (London): 219, 83.

Snapper, I. and Grunbaum, A. (1927) Biochem. Z.: 185, 223.

Soeldner, J.S. (1972) Arch. Surg.: 105, 683.

Thoren, L. (1964) Acta Chir. Scand. Suppl.: 325, 75.

Van Duyne, C.M. and Havel, R.J. (1959) Proc. Soc. Expt. Biol. Med.: 102, 599.

Williamson, D.H., and Hems, R. (1970) In: Essays in cell metabolism, p. 257. (Bartley, W., Kornberg, H.L. and Quayle, J.R., eds.) London: Wiley-Interscience.

Williams, D.H., Mellanby, J., and Krebs, H.A. (1962) Biochem. J.: 82, 90.

Williamson, J.R. and Krebs, H.A. (1961) Biochem. J.: 80, 540.

Wick, A.N. and Drury, D.R. (1941) J. Biol. Chem. 138, 129.

Woods, H.F., and Alberti, K.G.M.M. (1972) Lancet: 11, 1354.

Yeung, D. and Oliver, I.T. (1967) Biochem. J.: 105, 1229.

THE UTILIZATION OF XYLITOL, FRUCTOSE AND SORBITOL

Harald Förster

Biochemical and Physiological Institute

University of Frankfurt/Main

Carbohydrates represent an essential component of the caloric intake in intravenous nutrition. There has been experience with glucose, fructose and the two sugar alcohols, xylitol and sorbitol (Bassler, 1971; Mehnert, et al., 1970), glucose being the most important carbohydrate in parenteral nutrition. There are, however, certain conditions in which glucose utilization is impaired, such as diabetes mellitus or in the postoperative state and additional administration of insulin may be necessary.

In contrast to the dependence of glucose on insulin mediated transport into the peripheral cell, fructose, sorbitol and xylitol are hardly taken up by the peripheral tissue. Also in contrast to glucose, these sugars are metabolized primarily by the liver (Bassler, 1971; Forster, 1972) where they are converted to glucose and released into the blood or temporarily stored as glycogen. Thus fructose, sorbitol and xylitol contribute to the maintenance of blood glucose concentration by hepatic gluconeogenesis and probably reduce gluconeogenesis from other precursors such as protein. The hepatic utilization of fructose, sorbitol and xylitol causes some metabolic alterations, i.e., increased uric acid synthesis, increased lactate production and an increased lactate/pyruvate ratio.

Sorbitol is metabolized by the liver (Fig. 1) and oxidized to fructose by the enzyme sorbitol dehydrogenase in the course of which reaction NAD is reduced (Fig. 1). Fructose is then phosphorylated to form fructose-1-phosphate by hepatic phosphofructokinase in a reaction requiring ATP. Through the action of hepatic 1-phosphofructoaldolase, the fructose-1-phosphate yields dihydroxyacetonephosphate and glyceraldehyde. While dihydroxyacetonephosphate may enter the glycolytic pathway, glyceraldehyde is eventually phosphorylated by triokinase of oxidized or glyceric acid and subsequently phosphorylated by glyceric acid kinase (Hollman, 1961; Heinz et al., 1968), however, the fate of glyceraldehyde is not entirely clear.

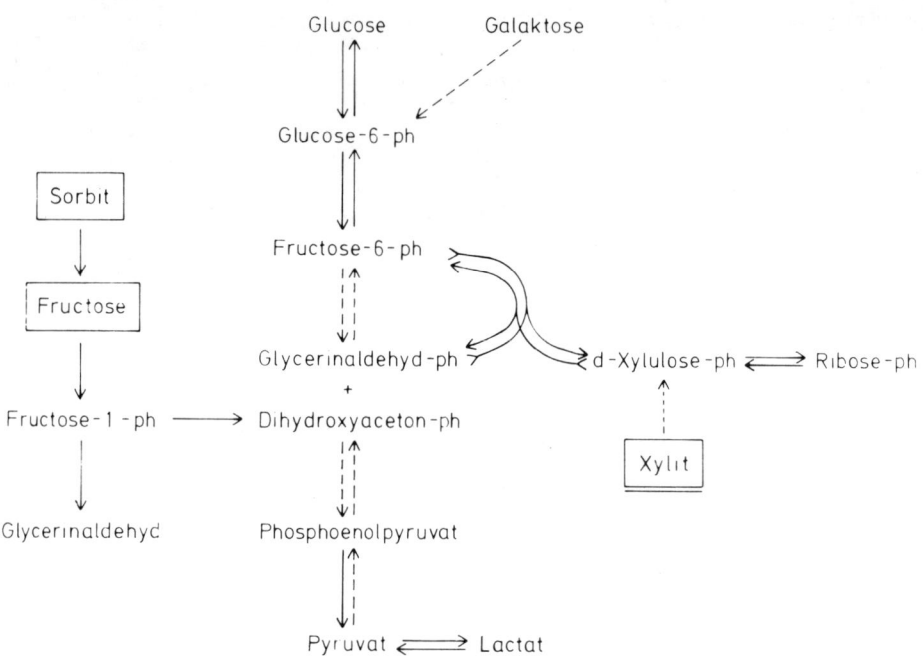

Fig. 1. Carbohydrate pathways.

Xylitol is oxidized to d-xylulose by the same enzyme which initiates sorbitol metabolism. The intramitochondrial xylitol dehydrogenase is not involved in the metabolism of exogenous xylitol (Hollman, 1961; Lang, 1971). D-xylulose is phosphorylated to d-xylulose-5-phosphate and subsequently converted to fructose-6-phosphate and glyceraldehyde-3-phosphate by means of the transaldolase and the transketolase reactions of the pentose phosphate cycle. Accordingly, xylitol is converted to glycolytic components or back to glucose. The metabolism of all three sugars is catalyzed by enzymes localized exclusively in the cytoplasmic compartment of the cell and the products of their metabolism are mainly channelled into the glycolytic pathway which is also in the cytoplasm.

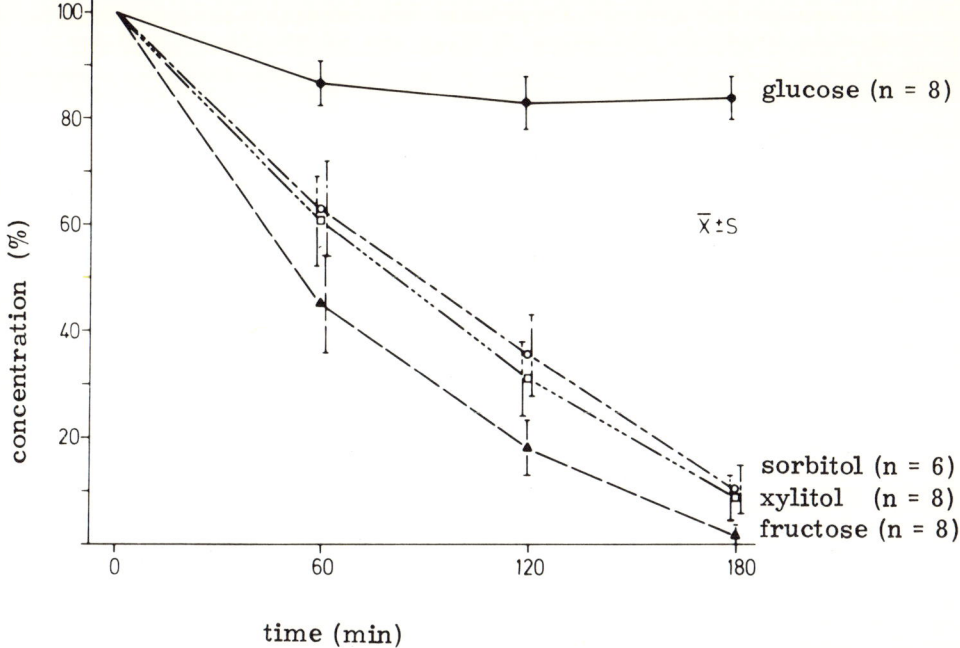

Fig. 2. Change in concentration of glucose (initial concentration 2.6 mM), xylitol (initial concentration 2.4 mM), sorbitol (initial concentration 2.0 mM) and fructose (initial concentration 2.0 mM) during perfusion of isolated rat liver.

As shown in Fig. 2, fructose as well as xylitol and sorbitol is taken up rapidly by the isolated perfused rat liver. However, glucose in high concentrations is obviously not metabolized by this organ preparation. We tried without success to stimulate glucose uptake. For example, insulin addition to the perfusion solution had no effect. The metabolic state of the perfused rat liver is probably similar to the postabsorptive situation; it does not take up glucose nor does it store glycogen.

On the other hand, glucose is formed very rapidly from fructose, sorbitol and xlitol in the perfused rat liver. During a three hour perfusion, 60-70% of added xylitol and 45-50% of fructose or sorbitol are converted to glucose. When fructose or sorbitol is perfused 20-25% of the sugars are converted to lactate, whereas only 5-10% of the added xylitol are converted to lactate (Fig. 3).

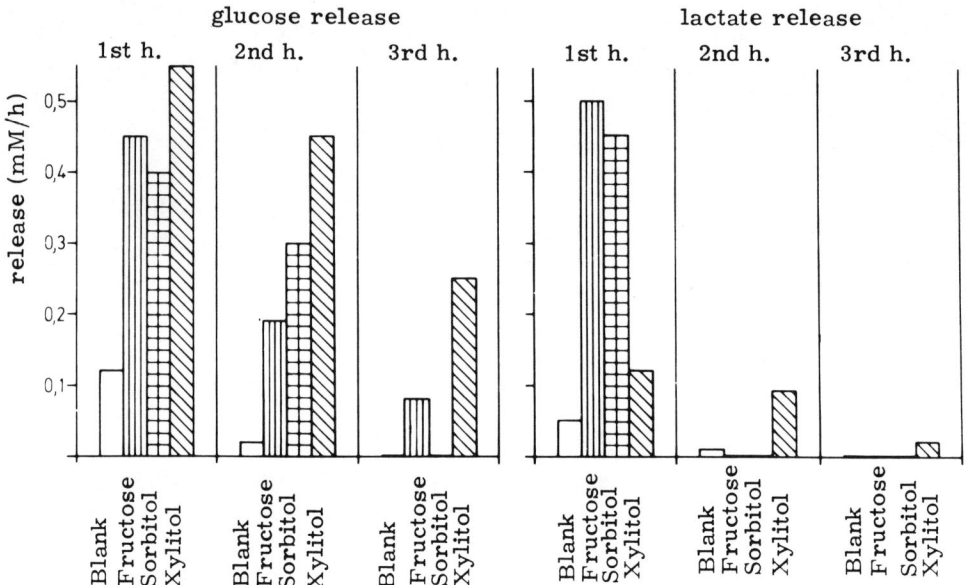

Fig. 3. Release of glucose and lactate by the isolated perfused rat liver following addition of 2.0 mM fructose, 2.0 mM sorbitol or 2.4 mM xylitol to basal medium.

During the metabolism of the polyalcohols, the cytoplasmic NAD is largely reduced, as is demonstrated by elevation of the lactate/pyruvate ratio (Fig. 4). Addition of fructose caused only minor changes in this ratio and glucose was without any effect. Despite these changes in cytoplasmic redox potential, there was extensive glucose production from the sugars (see Fig. 3).

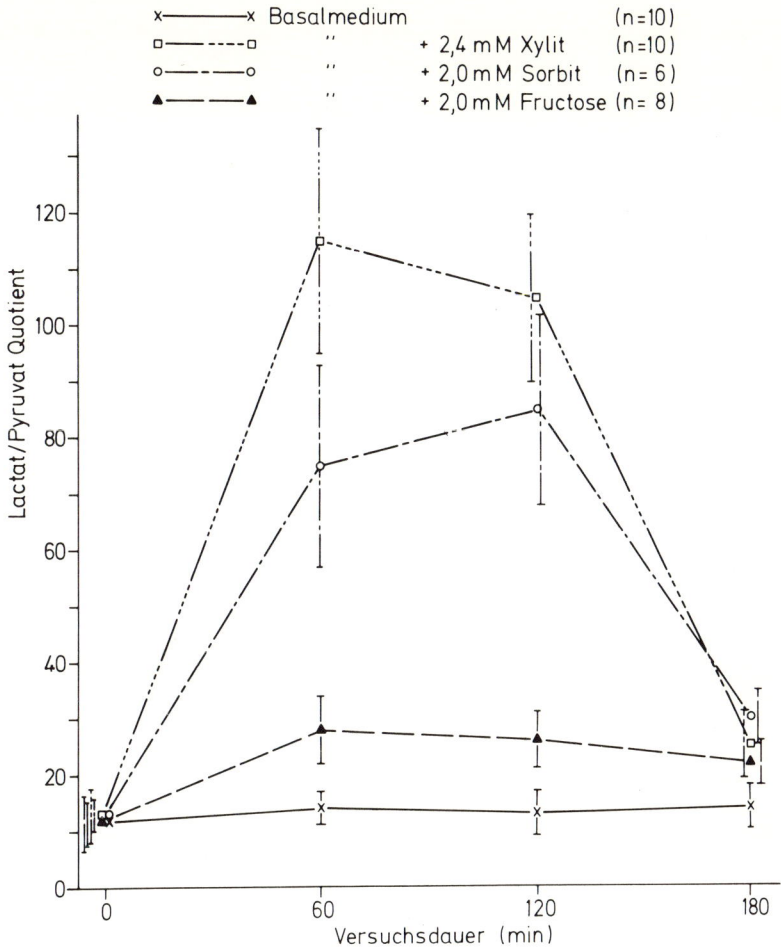

Fig. 4. Influence of fructose, sorbitol, xylitol and glucose on the lactate/pyruvate quotient during rat liver perfusion. x ± S.D.

The increased lactate formation during fructose metabolism is probably due to inhibition of diphosphofructose aldolase by accumulation of fructose-1-phosphate. This enzyme is involved in gluconeogenesis as well as in glucose breakdown and its inhibition would channel the metabolism of fructose to pyruvate and lactate. In fact, the lactate formation following xylitol administration was minimal, whereas fructose and sorbitol were nearly equally effective in this respect.

Glucose synthesized by the isolated liver during the fructose, sorbitol or xylitol perfusion is released into the circulation only under the experimental conditions of isolated organ perfusion. The inability of this preparation to store glycogen is one of the most prominent disadvantages of an otherwise very good experimental model. However, in whole animal studies, considerable glycogen deposition was demonstrated during intravenous infusion of the various carbohydrates. At a dose of 0.15 g per hour, i.e. 0.4 g per kg body weight, glycogen deposition over a two hour period accounted for as much as 50% of the infused carbohydrate (Forster, 1972). The actual glycogen synthesis from infused carbohydrate is lower. Using ^{14}C-glucose it has been shown that only 35-60% of the glycogen-^{14}C stems from the carbohydrates infused (Förster, et al., 1972). The remaining glycogen obviously derives from gluconeogenesis. When the infusion rate was doubled (Fig. 5), the glycogen concentration in the liver increased, particularly in those animals infused with fructose, sorbitol and especially xylitol and this occurred despite nearly normal blood glucose concentrations. The differences in glycogen deposition between animals receiving glucose infusions and animals receiving infusions of the other sugars are statistically significant as are the differences in the elevation of blood glucose concentration.

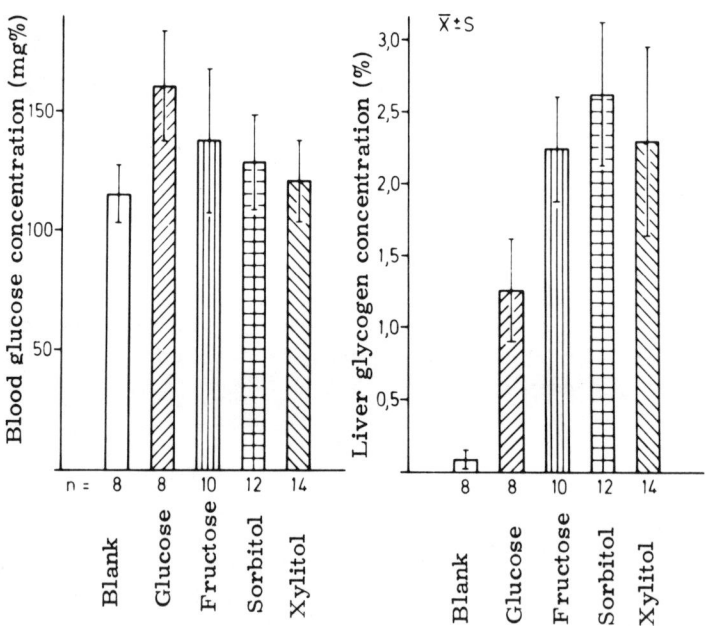

Fig. 5. Blood glucose concentration (2nd hour) and liver glycogen concentration following 2 h intravenous infusion of glucose, fructose, sorbitol and xylitol (0.3 gm/h) in rats.

THE UTILIZATION OF XYLITOL, FRUCTOSE AND SORBITOL

Glycogen formation in diabetic animals is observed only when fructose, sorbitol or xylitol is administered (Fig. 6). No glycogen is stored in the liver following glucose infusion. This was shown by Bassler and his co-workers (1963) in alloxan diabetic animals and we have confirmed these results using streptozotocin diabetic animals (Förster et al., 1972).

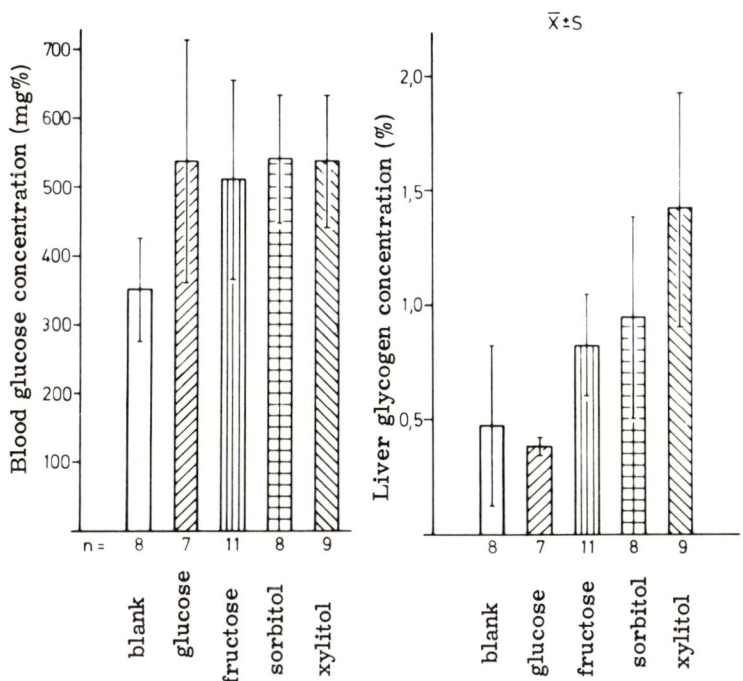

Fig. 6. Blood glucose concentration (2nd hour) and liver glycogen concentration following 2 h intravenous infusion of glucose, fructose, sorbitol, and xylitol (0.3 gm/h) in rats with streptozotocin diabetes.

In the diabetic animals, the increase in blood glucose concentration was highest after glucose infusion and was considerably less after infusion of the other substances.

Studies were also carried out with normal human volunteers. We have infused 1.5 g per kg body weight of a 20% solution of the various carbohydrates into male subjects. After a 15-20 minute infusion, changes in lactate, pyruvate, free fatty acids, uric acid and bilirubin concentrations in blood were followed (Forster et al., 1970; Forster, 1972). Except during infusion of glucose its blood concentration did not increase significantly. This is probably due to the slow transformation of the other carbohydrates to glucose, as well as to glycogen deposition (Fig. 3 and Fig. 5) and to other factors, such as a decrease in gluconeogenesis from protein. Other metabolic changes were similar to those found with the rat liver perfusion model.

As is shown in Figure 7, serum lactate concentration increased most after fructose infusion. But sorbitol was nearly as effective and glucose also caused a significant increase in blood lactate. However, confirming the liver perfusion studies, lactate increased only slightly after xylitol infusions. Xylitol infusion caused

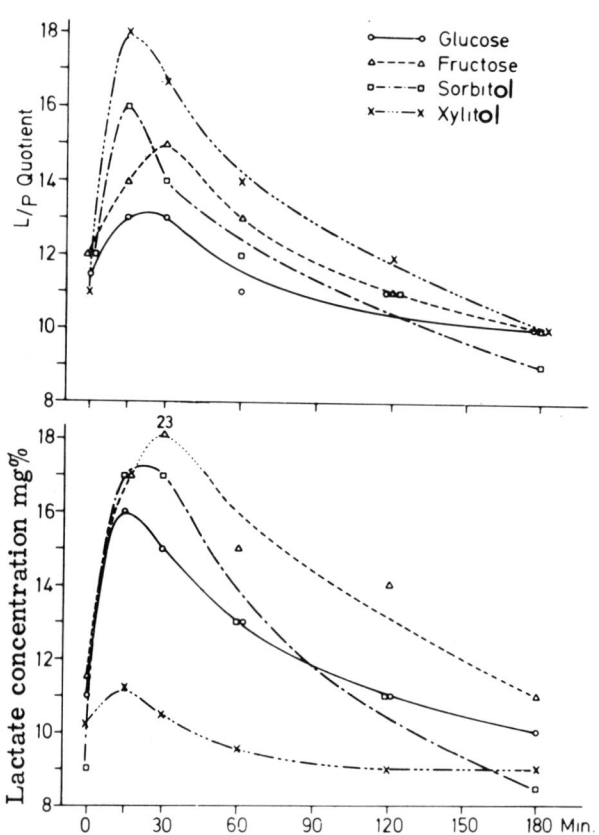

Fig. 7. Lactate/pyruvate quotient and lactate concentration in fasting voluntary subjects following intravenous infusion of glucose, fructose, sorbitol or xylitol (1.5 gm per kg BW).

the greatest change in the lactate/pyruvate ratio, however, glucose was also effective. These changes in lactate concentrations and in lactate/pyruvate ratio in human subjects were small in comparison to the liver perfusion studies. This is partially due to the fact that in the volunteers the estimations were performed with blood drawn from the cubital vein and not with blood from the hepatic veins. In contrast to isolated liver, normal volunteers showed significant changes with glucose and galactose administration as well as with the other carbohydrates.

We also studied the acute effect of reduction of cytoplasmic NAD on galactose metabolism by the liver. This is limited by the conversion of UDP-galactose to UDP-glucose by UDP-glucose epimerase. The coenzyme for this reaction which takes place in the cytoplasm of liver cells is probably NAD which must be kept in the reduced state to be effective. Neither xylitol nor sorbitol effectively inhibited galactose metabolism in healthy volunteers. These results demonstrate that the reduction in cytoplasmic NAD achieved by xylitol or by sorbitol does not influence metabolism in the same manner as ethanol. After ethanol administration gluconeogenesis (Lieber, 1967) as well as galactose metabolism (Forsander, 1966) is diminished. This is in contrast to the effects of the polyalcohols. Moreover, we have observed in our latest experiments that ethanol inhibits the conversion of fructose, xylitol and sorbitol to glucose in the perfused rat liver.

The administration of xylitol and fructose as well as sorbitol caused an increase in serum uric acid concentration (Fig. 8).

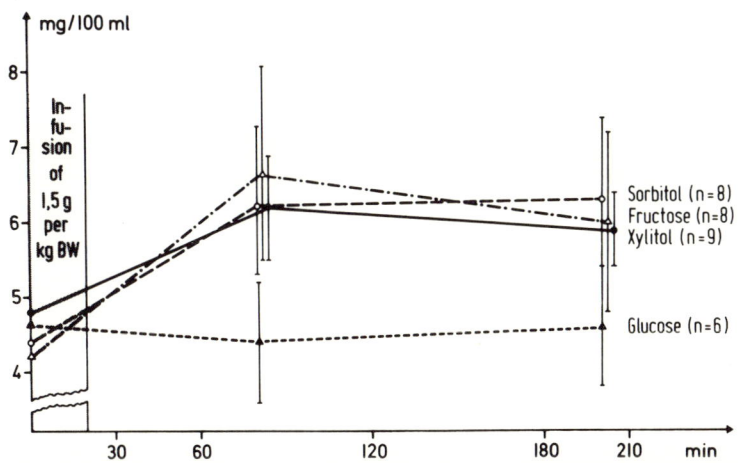

Fig. 8. Serum Uric Acid in Healthy Volunteers (Förster et al., 1970).

This is one of the so-called "adverse effects" drawn attention to by Schumer (1971) and by Edwards ad Thomas (1972). We observed this effect some years ago with fructose (Förster et al., 1967) and have also demonstrated similar findings with sorbitol and xylitol (Förster et al., 1970). We have also found increases in uric acid levels after oral administration of fructose, sorbitol, xylitol as well as sucrose (Förster et al., 1971). Omission of sucrose from the diet for a week caused a decrease in the concentration of uric acid. On the other hand, a daily oral intake of 200 gm sucrose, 100 gm fructose, or 100 gm xylitol effected an increase in serum uric acid concentration (Förster et al, 1972).

We demonstrated that rapid intravenous infusion of carbohydrate at a rate of 1.5 g per hour was followed by an increase in the serum bilirubin concentration (Fig. 9).

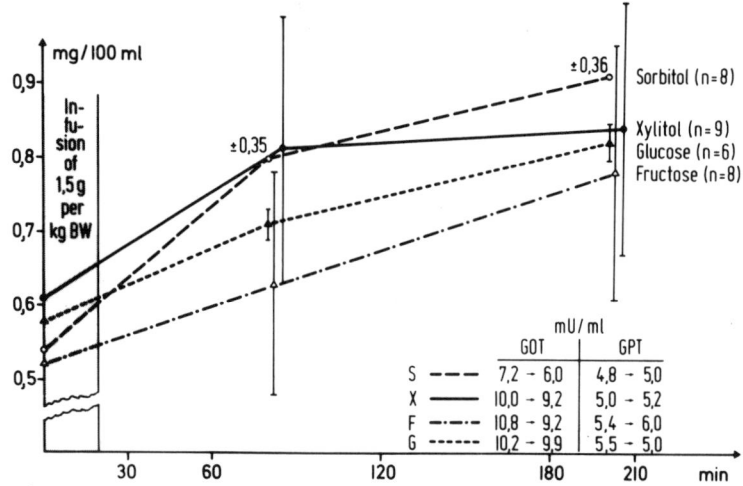

Fig. 9. Serum bilirubin in healthy volunteers (Förster et al., 1970).

The increase was identical whether glucose, fructose, xylitol or sorbitol was used (Förster et al., 1970).

Schumer (1971) additionally demonstrated an increase in serum enzyme levels which we have been unable to confirm. We have never found increased enzyme activities following intravenous carbohydrate infusions in human subjects or in the rat (Förster et al, 1970; Förster, 1972).

We also performed intravenous infusions of carbohydrates in rats for 72 hours. At the end of the infusion period, blood was drawn and SGOT, SGPT and bilirubin determinations (figure 10).

THE UTILIZATION OF XYLITOL, FRUCTOSE AND SORBITOL

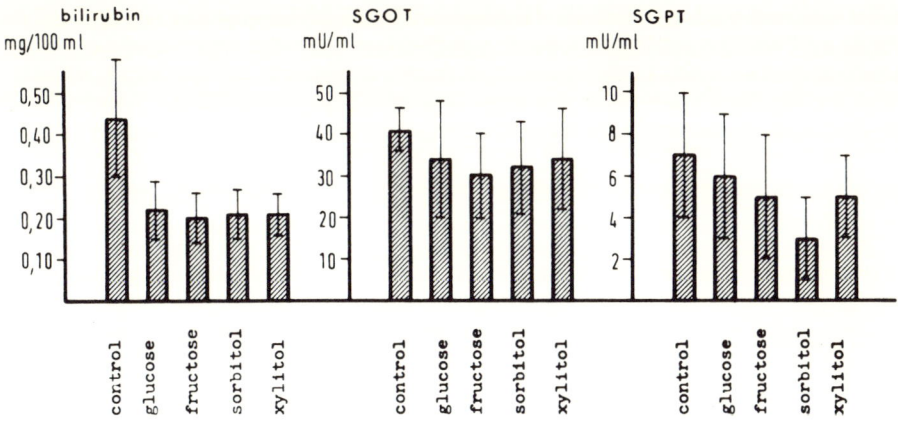

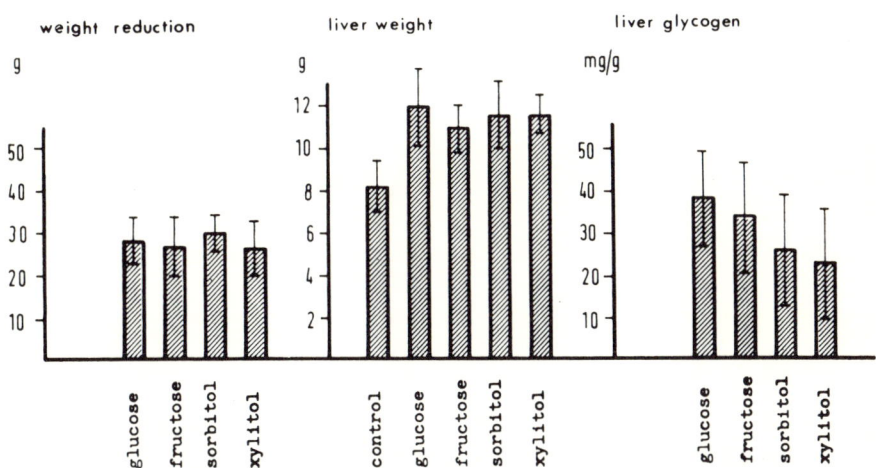

Fig. 10. Effect of 72 lasting intravenous infusions of some carbohydrates in rats on some parameters. D = 5.6 g/day.

The values were lower than in control animals. However, animals receiving continuous carbohydrate infusion showed an increase in liver glycogen (Förster et al., 1970). This was true despite the extreme stress of being immobilized in the cages during the infusion period. These findings are also inconsistent with any degree of hepatotoxicity caused by the infused carbohydrates. In neither group was there an increased incidence of death during the infusions.

To evaluate the benefit of the various carbohydrates, we looked at their metabolic effects under conditions of extreme hepatic gluconeogenesis. This involved the use of phlorizin diabetes first studied by Graham Lusk sixty years ago (1912). Phlorizin blocks renal tubular reabsorption of glucose, leads to renal glycosuria which results in gluconeogenesis. We administered phlorizin by means of continuous intravenous infusion with or without added carbohydrates.

The renal glucose loss in phlorizin diabetic rats was more than 1 g per 24 hours (Figure 11). Therefore, in excess of 2 g of protein had to be metabolized to glucose. If expressed as body weight, the glucose excretion amounted to 3-5 g per kg per day, and the nitrogen loss to 2-3 g per kg body weight per day.

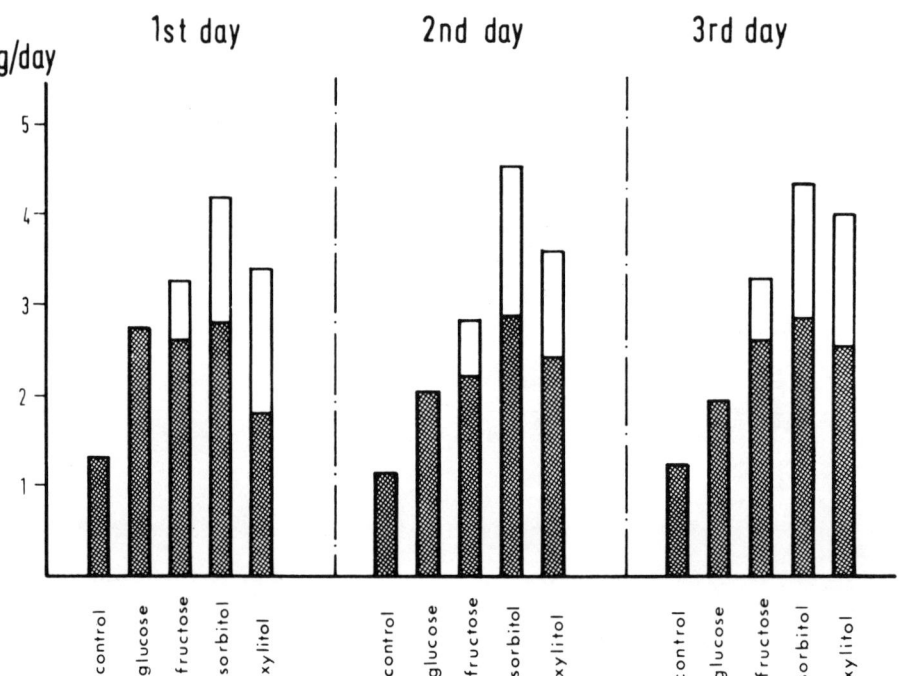

Figure 11. Sugar excretion following intravenous infusions of various carbohydrates (D = 5.6 g per day) and phlorizin (D = 0.01 g per day) in rats. Hatched bars = glucose excretion, open bars = fructose or polyol excretion.

As shown earlier by Lusk (1912), glucose excretion was

increased by the phlorizin treated rats when glucose or the other carbohydrates were infused. The glucose loss was greatest following sorbitol infusion and the smallest during xylitol infusion. However, even under these conditions the greater part of the infused substances was metabolized. Urine carbohydrate losses in most cases were less than 50%. Glucose and, to a somewhat lesser degree, fructose were best metabolized. After infusion of sorbitol, 5.6 g per day, approximately 1.5 g of sorbitol and 1.5 g of glucose were excreted by the kidneys. The renal excretion of urea was significantly reduced by any of these substances in the first two days of the experiment (Figure 12) suggesting a sparing effect on gluconeogenesis. On the last day of the infusion, no differences in the nitrogen excretion were observed.

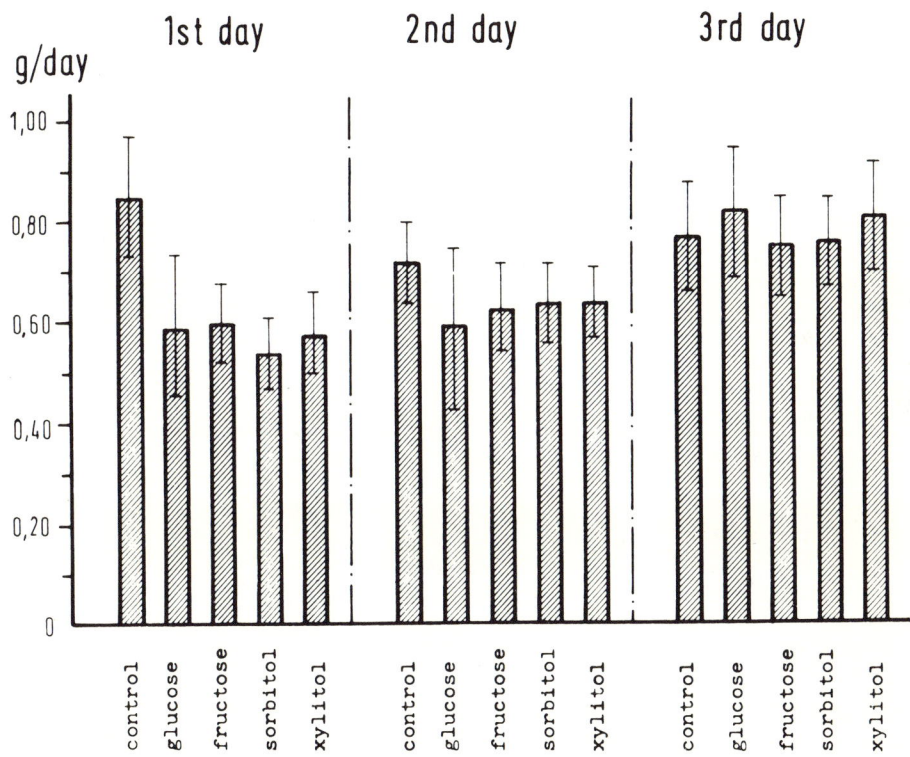

Fig. 12. Urea excretion during intravenous infusion of various carbohydrates (5.6 g per day) and phlorizin (0.01 g per day) $\bar{x} \pm$ S.D.

Interestingly, we made the unexpected observation that the blood glucose concentration rose during the initial period of the experiment. This elevated basal glucose level was caused by restraining the rats in the metabolic cages (Figure 13).

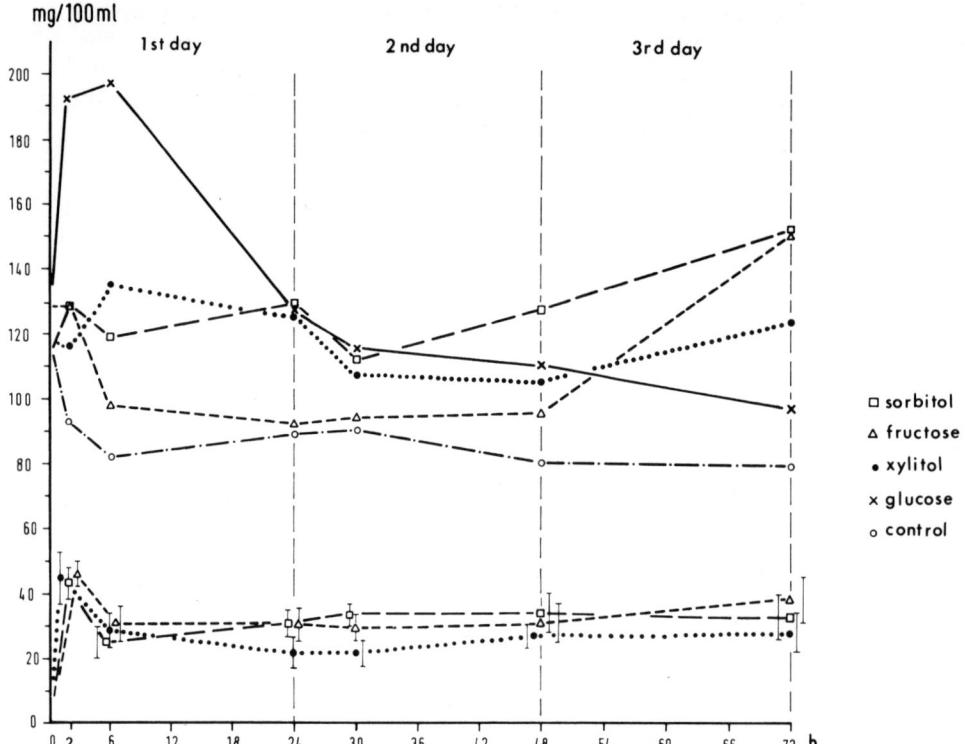

Fig. 13. Blood glucose concentration and substrate concentration during intravenous infusion of various carbohydrates (D = 5.6 g per day) and phlorizin (D = 0.03 g per day).

At the end of the experimental period, using sorbitol or fructose infusions, there was a rise in blood glucose concentration whereas the blood glucose concentrations during xylitol infusion was nearly identical with that of glucose infusions.

The blood concentrations of sorbitol, xylitol and fructose were nearly identical over the entire 72 hour infusion period. In contrast, renal fructose excretion was much less than observed for sorbitol. Xylitol excretion was intermediate. In human subjects xylitol and sorbitol were significantly less well utilized than was fructose when continuous infusion studies were performed (Berg et al., 1972; Keller and Froesch, 1972). On the other hand, the elimination rates of fructose, sorbitol and xylitol in human subjects were comparable to glucose (Bassler, 1971; Keller and Froesch, 1972).

CONCLUSION

The extensive investigations performed with human volunteers, experimental animals and the isolated liver have shown that fructose, sorbitol and xylitol can be effectively utilized by the liver. One of the main products of their metabolism is glucose which is continuously required by the brain, the blood cells and by the intestine. Therefore, hepatic glucose production from sorbitol, xylitol and fructose is not an artifact but under the conditions of administration contributes to the whole body glucose homeostasis. During short periods of fasting most of the glucose required is contributed by liver glycogenolysis. Under conditions when glycogen stores are depleted, glucose is formed from proteins via gluconeogenesis. In phlorizin diabetes in rats, the nitrogen sparing effect of glucose as well as by fructose, sorbitol and xylitol has been demonstrated.

In the postoperative state, diabetes, renal disease, as well as in other conditions, glucose utilization by the liver is disturbed. However, glucose-dependent tissues such as brain and blood must metabolize glucose. This need for glucose is met best by the administration of glucose precursors such as xylitol, sorbitol and fructose which are transformed into glucose by the liver even in the absence of insulin and have the additional advantage of reducing nitrogen catabolism. The administration of these carbohydrates also allows glycogen deposition in the livers of diabetic animals in whom glucose can not be utilized. Hyperglycemia is generally not seen following infusion of these substances.

In contrast to other authors, we have been unable to find severe adverse reactions by these substances. The rise in lactate concentration and the increase in the lactate/pyruvate ratio are common metabolic consequences of carbohydrate metabolism. The increase in uric acid and serum bilirubin concentrations has been reported as adverse reactions. However, the increase in uric acid concentration was also observed after oral sucrose intake. Rapid intravenous infusions of fructose, sorbitol and xylitol at high dosages of 1.5 g per kg were followed by an increase in the serum bilirubin concentration. Glucose, however, caused the same degree of increase. In extensive studies with experimental animals, we have not found any adverse reactions caused by infusions of sorbitol, xylitol and fructose. In contrast to the studies in dogs, hypoglycemic reactions following the infusion of xylitol are not observed in humans or in rats.

REFERENCES

Bässler, K.H. and Heesen, D. Klin. Wschr., 41, 595, 1963.

Bässler, K.H. Z. Ernährungsw., Suppl., 10, 57, 1971.

Berg, G., Bickel, H. and Matzkies, F. 9th International Congress of Nutrition, 1972.

Edwards, R.G. and Edwards, J.B. Technicon Symposion, Frankfurt, Germany, 1972.

Forsander, O.A. Scand. J. Clin. Lab. Invest., 18, Suppl., 92, 143, 1966.

Förster, H., Boecker, S., Ziege, M. Med. u. Ernährung, 13, 193, 1972.

Förster, H., Mehnert, H. and Alhough, I. Klin. Wschr., 45, 436, 1967.

Förster, H., Meyer, E. and Ziege, M. Klin. Wschr., 48, 878, 1970.

Förster, H. and Ziege, M. Z. Ernährungsw., 10, 524, 1971.

Förster, H. Med. u. Ernährung, 13, 7, 1972.

Förster, H., Lerche, D. and Hoos, I. Kongr. Deutsche Diabetes Ges., 1972, Abstract 52.

Förster, H. Z. Ernährungsw., 11, 227, 1972.

Förster, H. Haslbeck, M., Buchele, H. and Mehnert, H. Z. Ernährungsw. (in press)

Heinz, F., Lamprecht, W.L. and Kirsch, J. J. Clin. Invest., 47, 1826, 1968.

Hollmann, S. Nicht glykolytische Stoffwechselwege der Glucose, Thieme, Stuttgart, 1961.

Keller, U. and Froesch, E.R. Schweiz. med. Wschr., 102, 1017, 1972.

Lang, K. Klin. Wschr., 49, 233, 1971.

Lieber, C.S. Fed. Proc., 26, 1443, 1967.

Lusk, G. Ergebn. Physiol., 12, 315, 1912.

Mehnert, H., Förster, H., Geser, C.A., Haslbeck, M. and Dehmel,
    K.H. in Parenteral Nutrition (eds. C.H. Meng and D.H. Law),
    Charles C. Thomas Publishers, Springfield, 1970, p. 112.

Schumer, W. Metabolism, 20, 345, 1971.

FATTY ACID OXIDATION DURING DEVELOPMENT

Joseph B. Warshaw

Departments of Pediatrics and Obstetrics, Yale University

School of Medicine, New Haven, Connecticut

There has been considerable interest in the use of fatty acids as a source of calories in parenteral nutrition. From the standpoint of energy economy fatty acids offer a distinct caloric advantage over carbohydrates and proteins in that they provide nine calories per gram administered in contrast to the approximately four calories yielded by carbohydrates and amino acids. Thus, from the weight volume standpoint, the administration of fatty acids in parenteral nutrition can provide patients with significant nutritional increments. In addition, fatty acids are the preferred energy fuel of a number of tissues including heart (Shipp et al., 1961), renal cortex, and small intestinal mucosa (Hulsmann, 1971) so that their availability and administration is likely to spare both carbohydrate and protein catabolism by those and other tissues. The capacity of mammalian tissues to utilize fatty acids as an energy fuel develops after birth as part of the general metabolic adaptation of the mammalian newborn to the postnatal environment. It has been long recognized that energy requirements during embryonic and fetal development are provided primarily by carbohydrates (Needham and Lehman, 1937).

In this presentation the pathways supporting fatty acid oxidation will be reviewed with particular emphasis placed on the development of fatty acid oxidation during the postnatal period. Such information is essential if potential hazards as well as benefits of lipid administration are to be rationally understood and their use put in the best perspective.

Cellular Oxidation of Long-Chain Fatty Acids

Fatty acids presented to the tissues may derive from several sources. They may be made available as a result of intestinal absorption, by hydrolysis of circulating triglycerides or released from adipose tissue stores. Fatty acids and monoglycerides of greater than 10 carbon chain length are first transported into the intestinal epithelial cell where they are re-esterified, transported into the intestinal lymphatics and ultimately to the thoracic duct and liver, as chylomicra. On the other hand, fatty acids and monoglycerides of less than 10 carbon chain length may be absorbed into the portal venous system and transported directly to the liver. The hepatic fatty acids are repackaged in the form of triglycerides and lipoproteins from which fatty acids can be released at their sites of tissue metabolism by the action of lipoprotein lipase which can be stimulated by heparin (Schalch and Kipnis, 1965).

Long chain fatty acids derived from the adipose tissue are released through the action of the adipose tissue lipase (Vaughan et al., 1964) and the free fatty acids released from the adipose tissue are transported in the circulation bound to albumin. As it well known, lipase activity increases in response to β-adrenergic stimulation via the activation of adenyl cyclase and the second messenger, cyclic AMP (Butcher et al., 1972). On the other hand, insulin exerts an anti-lipolytic effect (Butcher et al., 1968).

Figure 1 depicts the sequence of events associated with the cellular uptake of fatty acids and fatty acid oxidation. Fatty acid uptake by the cell takes place by mechanisms not completely understood. Fatty acid analogues such as bromo-palmitate have been reported to inhibit the intracellular accumulation of long-chain fatty acids (Mahadevan and Saver, 1971) suggesting that their transport may depend on specific transport systems. Once inside the cell, the fatty acids may be bound to anionic binding proteins such as described by Arias and others (Ockner et al., 1972; Mishkin et al., 1972). They have demonstrated the presence of such proteins in liver, the so-called Y and Z proteins which have high affinity for bilirubin. A similar binding role for intracellular fatty acids has been attributed to the so-called Z protein. The binding proteins themselves may be important in the uptake of the fatty acids at the cell membrane.

In the case of the hepatocyte, the binding proteins presumably deliver long chain fatty acids to sites on the endoplasmic reticulum where fatty acids are activated to their coenzyme A esters. Three distinct fatty acid activating enzymes (acyl-CoA synthetases) have been identified (Aas, 1971; Aas and Daae, 1971). Fatty acids of greater than 12 carbon chain length are activated to their corresponding coenzyme A ester by palmityl-CoA synthetase. Similarly,

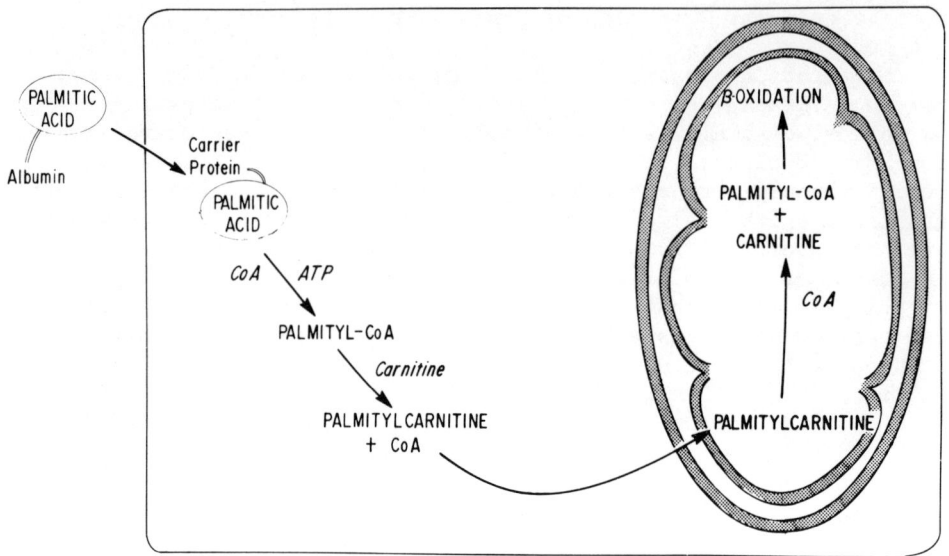

Fig. 1. Scheme showing cellular uptake, activation and mitochondrial transport of palmitic acid.

medium chain fatty acids, from 6 to 12 carbons chain length, are activated by medium chain activating enzymes and fatty acids of less than 6 carbon chain length are activated by a short chain fatty acid activating enzyme, acetyl CoA synthetase. Fatty acids of less than 10 carbon chain length freely permeate the mitochondrial membrane and can be activated to their corresponding coenzyme A esters in the mitochondrion itself. However, longer chain fatty acids permeate the mitochondrial membrane very slowly and the inner mitochondrial membrane is a barrier to the transport of coenzyme A esters of all fatty acids regardless of their chain length (Fritz and Marquis, 1965; Bremer, 1967). The transport and subsequent oxidation of CoA esters of fatty acid is dependent on the formation of acylcarnitine transport intermediates which allow the fatty acids to gain entry to the sites of mitochondrial oxidation. The synthesis

of acylcarnitine esters of fatty acids from the corresponding
coenzyme A esters is catalyzed by a group of enzymes known as
acylcarnitine transferases. Again, there is a specific long chain
acylcarnitine transferase for fatty acids of greater than 10 carbon
chain length (palmitylcarnitine transferase). A short chain enzyme,
acetylcarnitine transferase, catalyzes the formation of short chain
acylcarnitine esters of less than 10 carbon chain length and recent
evidence suggests the presence of a medium chain acylcarnitine
transferase.

Once the acylcarnitine esters of the fatty acids gain entry
into the mitochondria they are reconverted to their corresponding
coenzyme A esters and β oxidation proceeds. The reaction sequence
for these transitions is shown below.

$$\text{Palmityl-CoA} + \text{carnitine} \quad \text{palmitylcarnitine} + \text{CoA} \quad (1)$$

$$\text{Palmitylcarnitine} + \text{CoA} \quad \text{palmityl-CoA} + \text{carnitine} \quad (2)$$

The initial formation of palmitylcarnitine occurs external to
the membrane barrier to palmityl-CoA (reaction 1). Once transport
has been effected, the palmitylcarnitine is reconverted to the
coenzyme A ester (reaction 2). The reaction catalyzed by palmitylcarnitine transferase is an equilibrium reaction so that under
normal circumstances if oxidation is taking place the conversion
of extra mitochondrial palmityl-CoA to palmityl carnitine does
not become rate limiting for fatty acid oxidation. However, the
situation during development is quite different and the generation
of palmitylcarnitine external to the mitochondrial membrane barrier
may impose constraints on fatty acid oxidation.

Because of the importance of these enzyme activities for
active fatty acid oxidation, our laboratory has carried out investigations of fatty acid activation, acylcarnitine transferase activity and fatty acid oxidation itself in a variety of mammalian
sepcies during development.

Fatty Acid Oxidation During Development

As already noted above, the developing mammal is highly dependent on transplacentally derived glucose to provide its energy
needs. After birth, as the animal begins to adapt to a more efficient oxidative metabolism, tissues such as liver and heart show
an increase in both mitochondrial number and in specific mitochondrial enzymes such as cytochrome oxidase and various enzymes of
the citric acid cycle important for oxidative phosphorylation. In
the case of the developing heart, we have previously shown (Warshaw,
1972) that cytochrome oxidase activity of newborn rat heart homo-

genates increases two-fold during the first ten days of life. On the other hand, the oxidation of glutamate-malate shows a manyfold increase. Cytochrome oxidase has been used as an index of mitochondrial number. The much larger increase observed in the case of glutamate-malate oxidation reflects the fact that specific mitochondrial activities may also increase postnatally independent of the increase in mitochondrial number. Other investigators have made similar observations. For example, Hommes and Richter (1969) showed a six-fold increase in the number of rat liver mitochondria during the first month of development.

With respect to fatty acid oxidation itself, Breuer (Breuer et al., 1969), some years ago, showed that free fatty acids were not extracted from the coronary arteries of newborn puppies. Wittles and Bressler (1965) reported that the preferential oxidation of glucose in contrast to fatty acids by newborn rat heart homogenates was due to low levels of carnitine and decreased activity of the enzyme palmitylcarnitine transferase. Our own laboratory has carried out investigations of fatty acid oxidation by the developing bovine heart (Warshaw and Terry, 1970). It was demonstrated that acylcarnitine esters of fatty acids could be oxidized efficiently by fetal bovine heart mitochondria. However, although bovine fetal mitochondria were capable of efficient oxidative phosphorylation with acylcarnitine substrates, they did have an impairment of carnitine supported palmityl-CoA oxidation. This indicated to us that a deficiency in a compartment of palmitylcarnitine transferase external to the mitochondrial membrane barrier to palmityl-CoA was rate limiting for fatty acid oxidation by fetal bovine heart mitochondria. However, an internal compartment of palmitylcarnitine transferase could reconvert exogenous palmitylcarnitine to palmityl-CoA once transport had been effected. In Figure 2 is shown a comparison of ^{14}C palmityl-^{14}C-CoA + carnitine oxidation by isolated fetal and calf heart mitochondria. $^{14}CO_2$ production by the fetal mitochondria is significantly less than that observed with post-natal material.

Investigations of fatty acid activation, acyl transfer and oxidation were also carried out in the developing rat (Bremer, 1967). Palmityl-CoA synthetase activity of both heart and liver showed a rapid increase during the first days of life (Fig. 3). The short chain activating enzyme, acetyl-CoA synthetase, showed a similar developmental increment. Palmitylcarnitine transferase activity of developing rat liver and heart homogenates increased from negligible levels at the time of birth to adult levels by thirty days of age (Fig. 4). These changes were associated with a large increase in the overall rate of fatty acid oxidation in the rat (Fig. 5). Thus, both fatty acid activation and acylcarnitine transferase activity appear to be of considerable importance for the development of fatty acid oxidation in the rat. Augenfeldt

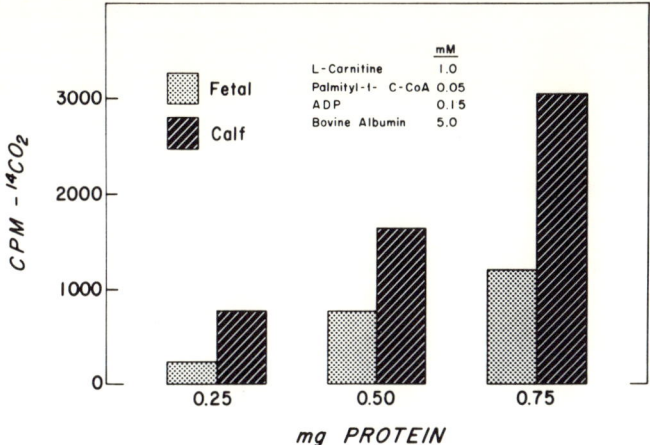

Fig. 2. Comparison of palmityl-CoA plus carnitine oxidation by isolated fetal and calf heart mitochondria.

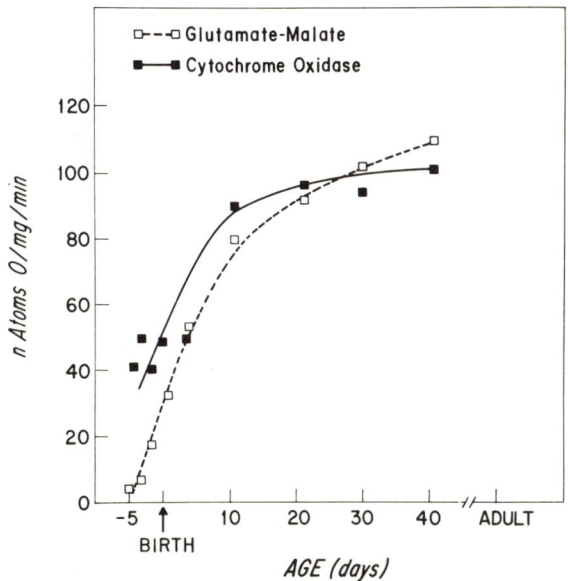

Fig. 3. Palmityl-CoA synthetase activity of developing rat liver. Activity is expressed as nanomloes of palmityl-CoA produced per min per mg protein (Warshaw, 1972).

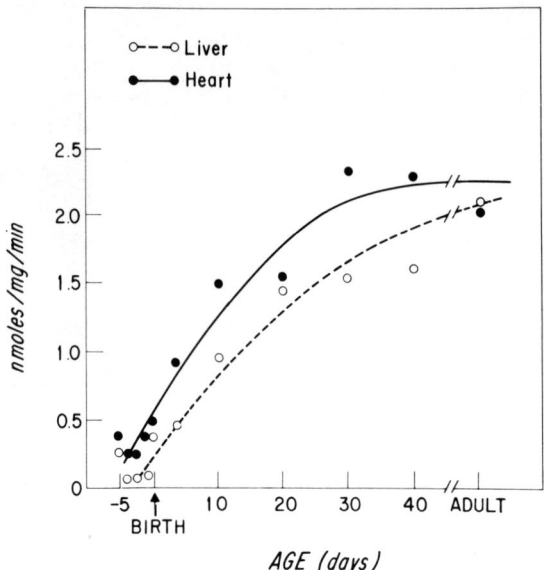

Fig. 4. Palmitylcarnitine transferase activity of developing rat heart and liver. Activity is expressed as nanomoles of palmitylcarnitine produced/min/mg protein (Warshaw, 1972).

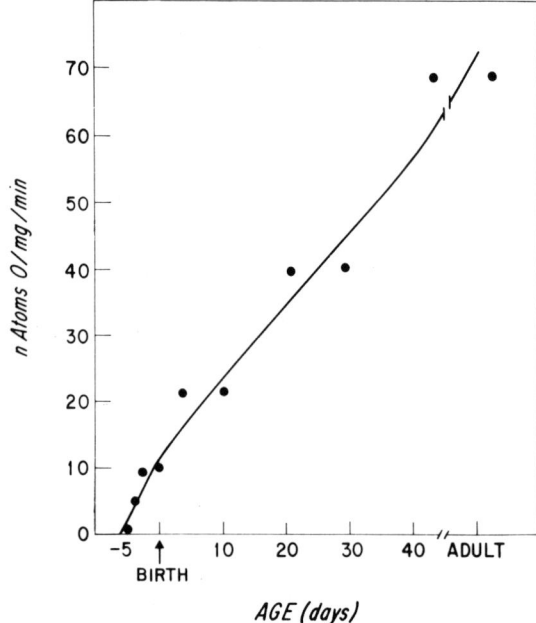

Fig. 5. Oxidation of palmityl-CoA plus carnitine by rat heart homogenates. Activity is expressed as nanoatoms oxygen utilized/min/mg protein (Warshaw, 1972).

and Fritz (1970) and Lockwood and Bailey (1970) also found that
the development of palmitylcarnitine transferase activity in the
developing liver paralleled a post-natal increase in fatty acid
oxidation.

As noted above, fatty acids are the preferred substrate of
the small intestine (Hulsmann, 1971). Investigations carried out
in our laboratory have demonstrated that long chain fatty acid
oxidation by homogenates of intestinal mucosa increases during the
post-natal period. The intestine, like other fetal organs, probably
oxidizes glucose preferentially during fetal development. However,
because of the high energy demands placed on the gut during the
suckling, fatty acid oxidation rapidly becomes an important source
of energy production. As shown in Table I, fatty acid oxidation
by homogenates of intestinal mucosa was much lower in the fetal
material as compared with two week old and two month old calves.

Table 1: Fatty acid oxidation by small intestinal homogenates

Substrate	Fetus* 6 months	Calf 2 weeks	Calf 2 months
	nmoles oxidized/mg/hr		
capric acid (C-16)	1.5	5.05	13.2
palmitic acid (C-16)	2.5	7.1	9.1

* Gestation in the cow is nine months.

Cytochrome oxidase activity of the two-month old calf intestine
was approximately twice that of fetal material indicating that the
number of mitochondria as well as the capacity for fatty acid
oxidation also increased. These observation are also similar to
those observed in the case of the developing heart and emphasize
the increasing postnatal importance on aerobic oxidations.

Thus, it appears that the increased availability of oxygen
during the postnatal period is associated with an enhanced capacity
for oxidation of long chain fatty acids. Indeed, situations in
which there is persistent hypoxia may be similar to the prenatal
condition and impose limitations on fatty acid oxidation. This

has been shown experimentally by Kjekshus and Mjøs (1972) who investigated the effects of intravenous fat emulsions on cardiac performance in hypoxic dogs. They demonstrated that under conditions of hypoxia there was a significant decrease in myocardial contractility. It is also likely that under conditions in which there is impaired oxygenation and oxidation fatty acids would accumulate in liver and predispose the individual to the pathologic manifestations of fatty liver. Indeed, Baum (1969) has shown that the release of fatty acids from adipose tissue is markedly impaired during acute hypoxia so that the organism appears to have a natural mechanism for protection against increases in fatty acids during hypoxic situations.

In the case of hyperalimentation, it would appear to be prudent to avoid the use of intravenous fat emulsions in situations associated with impaired oxygenation. Certainly, more information is necessary concerning fatty acid utilization in different pathological states prior to wide application of fat emulsions to newborn infants.

Acknowledgments

Work described here was supported by United States Public Health Service Grant No. HD 03610.

References

Aas, M. Biochim. Biophys. Acta, 231, 32-47, 1971.

Aas, M. and Daae, N. Biochim. Biophys. Acta, 239, 208-217, 1971.

Augenfeldt, J. and Fritz, I. Can. J. Biochem., 48, 228-234, 1970.

Baum, D. J. Pharmacol. Exptl. Therap., 169, 87, 1969.

Bremer, J. in Cellular Compartmentalization and Control of Fatty Acid Metabolism. F.C. Gran, editor. Academic Press, Inc., New York, 1967, p. 65.

Breuer, E., Barta, E., Zlatos, L. and Pappova, E. Biol. Neonat., 12, 54, 1969.

Butcher, R.W., Baird, C.E. and Sutherland, E.W. J. Biol. Chem., 243, 1705, 1968.

Butcher, R.W., Robison, G.A. and Sutherland, E.W. in Biochemical

Actions of Hormones, Vol. II. G. Litwack, editor. Academic Press, Inc., New York, 1972.

Butcher, R.W., Baird, C.E. and Sutherland, E.W. J. Biol. Chem., 243, 1705, 1968.

Fritz, I.B. and Marquis, N.R. Proc. Nat. Acad. Sci., U.S.A., 54, 1226, 1965.

Hommes, F.A. and Richters, A.R. Biol. Neonatorum, 14, 359-364, 1969.

Hulsmann, W.C. FEBS Letters, 17, 35, 1971.

Kjekshus, J.K. and Mjøs, O.D. J. Clin. Invest., 51, 1767, 1972.

Lockwood, E.A. and Bailer, E. Biochem. J., 120, 49, 1970.

Mahadevan, S. and Saver, F. J. Biol. Chem., 246, 5862, 1971.

Mishkin, S., Stein, L., Gatmaitan, Z. and Arias, I.M. Biochem. Biophys. Res. Comm., 47, 997, 1972.

Needham, J. and Lehman, H. Biochem. J., 31, 1210, 1937.

Ockner, R.K., Manning, J.A., Poppenhausen, R.B. and Ho, W.K.L. Science, 177, 56, 1972.

Schalch, D.S., Kipnis, D.M. J. Clin. Invest., 44, 2010-2020, 1965.

Shipp, J.C., Opie, L.H. and Challoner, D.R. Nature, 189, 1018, 1961.

Vaughan, M., Berger, J.E. and Steinberg, D. J. Biol. Chem., 239, 401-409, 1964.

Warshaw, J.B. Develop. Biol., 28, 537, 1972.

Warshaw, J.B. and Terry, M.L. J. Cell. Biol., 44, 354, 1970.

Wittles, B. and Bressler, R. J. Clin. Invest., 44, 1630, 1965.

UTILIZATION AND TOLERANCE OF INTRAVENOUS FAT EMULSIONS

Robert P. Geyer

Department of Nutrition, Harvard School of Public Health

Boston, Massachusetts

Emulsions of fat have been the subject of much research in the field of parenteral nutrition because they offer a concentrated source of calories in a moderate volume without the complications of hypertonic solutions. It is generally accepted that a suitable emulsion would be highly desirable in those instances where caloric requirements are high and excessive amounts of fluid must be avoided. Much of the effort in this area during the past three or four decades has been devoted to proving intravenously administered fat is utilized, is efficacious, and is devoid of serious reactions when properly formulated (Geyer, 1960; Wretlind, 1972; Geyer, 1970). This presentation is concerned with the utilization and tolerance of parenteral fat emulsions.

In principal, fat emulsions are simply minute fat droplets well dispersed in an aqueous medium. To aid in the preparation and stabilization emulsifiers such as phospholipids are employed. Although a number of procedures have been used to disperse the triglyceride, high pressure homogenization has proved to be the most satisfactory, especially for large quantities. The products have particle diameters of approximately one micron and below, and withstand sterilization in a rotating autoclave. Stability is sufficient to allow shipment anywhere in the world and storage can be for as long as a year or more. Oxidative and hydrolytic changes are minimal.

Since the purpose of these emulsions is to furnish calories, and from 10 to 20% fat (w/v) is usually employed, the fat of the triglyceride is obviously of prime interest. On the other hand, even though the emulsifying agent is present in much lower concentration (usually 1 to 2%), its fate is also of concern. Depending on its chemical character it may be metabolized, as in the case of lecithin, or excreted, as is the synthetic polyol, Pluronic F68 (Wyandotte Chemical Corp., Wyandotte, Michigan). Injection of the emulsified fat particles causes a lipemia, the degree of which depends upon the rates of administration and utilization. The disappearance of Intralipid (Vitrum, Stockholm, Sweden) (Hallberg, 1965a and Hallberg, 1965b) from the blood has been shown to follow first order kinetics when the concentration is below maximum. Above this concentration the rate is constant and independent of actual concentration. In the case of Intralipid there is good agreement between the kinetics of removal of the artificial fat droplets and normal chylomicrons whether compared in the presence of heparin-induced lipoprotein lipase or not (Hallberg, 1965a; Hallberg, 1965b; Boberg and Carlson, 1964). The size distribution of Intralipid particles is also quite similar to that of chylomicrons (Schoefl, 1968; Fraser and Hakansson, in press).

Intravenously administered triglycerides can be hydrolyzed by means of lipoprotein lipase present in the blood. Entrance of the intact triglyceride into cells and its subsequent hydrolysis by intracellular enzymes has not been demonstrated to date. Stoffel and coworkers have presented evidence that intact phospholipid can be taken up by many different cells when given intravenously (LeKim, Betzing and Stoffel, 1972). The products of lipase action, unesterified fatty acid, and intact monoglyceride, are known to be taken up by cells and metabolized (Lynch and Geyer, 1972). There are a number of indications that the intravenously injected fat is utilized (Geyer, 1960; Wretlind, 1972; Geyer, 1970) and these are summarized in Table 1.

Table 1: Observations supporting the concept that parenteral emulsions are utilized.

1) The fat particles quickly leave the bloodstream.
2) Little accumulation of lipid occurs in the tissues.
3) Essentially no lipid is lost via excretion.
4) Body weight responses are favorable.
5) RQ values shift towards fat oxidation.
6) Blood ketone values rise.
7) Labeled triglyceride is rapidly metabolized.
8) Fat can furnish an important percentage of calories in complete parenteral nutrition.

The rapid removal from the blood stream, the lack of significant losses through excretion, and the lack of accumulation of the injected fat in tissues even after large daily doses, furnish indirect evidence of utilization.

More direct evidence of utilization is the favorable effect fat emulsion administration has on body weight, the shift in respiratory quotient towards fat oxidation, and a rise in the concentration of blood ketones. Absolute evidence for the metabolism of the emulsified triglyceride is given by the conversion of ^{14}C-labeled fat to ^{14}CO$_2$ and ^{14}C-phospholipids following its intravenous injection. Finally, the fact that when fat emulsions comprise an essential part of the energy source in complete parenteral nutrition, growth and development occur, shows in a very practical manner that such emulsions are indeed utilized (Hallberg et al., 1970 and Hakansson et al., 1967).

It is of interest to compare the early stages of the metabolism of parenteral nutrients. Glucose and amino acids can enter oxidative and other pathways with little prior alteration in their structure. On the other hand, the fat particles in Intralipid must undergo a number of enzymatic reactions prior to the entry of the component parts into comparable metabolic pathways. This is illustrated in Fig. 1.

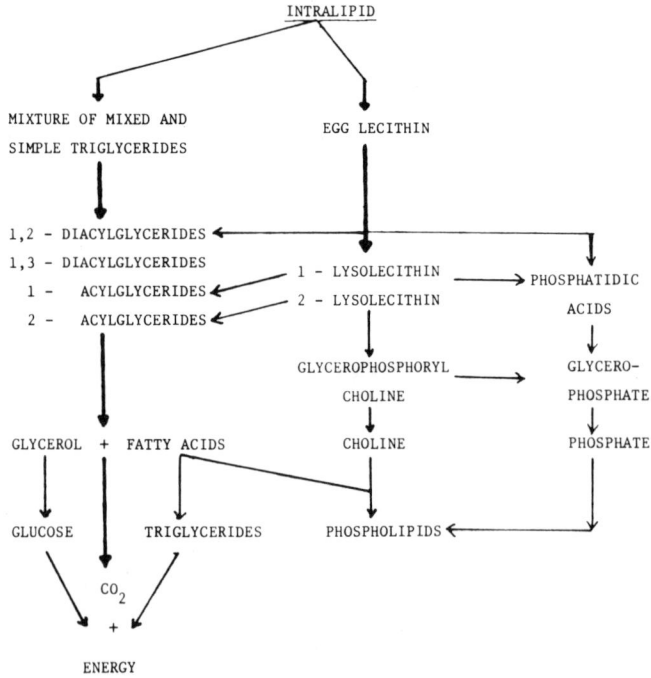

Fig. 1. Multiplicity of pathways involved in Intralipid metabolism.

The egg lecithin is converted to lysolecithin and phosphatidic acid and finally to choline, glycerol, fatty acids and phosphate. Triglyceride must be converted to glycerol and fatty acids before entering into energy-yielding pathways. The formation of partial acylglycerides affords a number of possibilities for conversion to other lipids. Obviously the expectation when giving parenteral fat emulsion is that the fat will be converted to CO_2 and H_2O and yield energy in so doing. However, the contribution of the administered lipids to the lipids of the recipient should not be overlooked. For example, supplying essential fatty acids and thereby prostaglandin precursors would be an important ancillary reason for giving parenteral fat emulsions. Coats and Collins (1969) have demonstrated linoleic acid deficiency in man and its alleviation with parenteral fat.

Should the need arise in the future for emulsions containing any specific lipids it is likely that this requirement can be met. It is to be noted that the main ingredients of fat emulsions which have been tested in humans are not chemically pure, as for example when egg lecithin and soybean oil are used. Again in the future, it may become desirable to use pure or even synthetically produced lipids; however, at present costs alone forestall their use.

As stated earlier, studies with radioactive glycerides have shown good utilization of intravenously injected emulsified lipids. Table 2 gives comparisons between the oxidation of intravenously administered doses of $1-^{14}C$-triolein, $1-^{14}C$-oleic acid and $1-^{14}C$-ethyl alcohol by the rat.

Table 2: Conversion of several parenterally administered radioactive nutrients to $^{14}CO_2$.

Nutrient	%
$1-^{14}C$ -Triolein	
Emulsion 1	63.9
Emulsion 2	62.0
$1-^{14}C$ -Oleic Acid	76.4
$1-^{14}C$ -Ethyl Alcohol	96.1

24-Hour collection from rats. All compounds given on identical millimole carbon basis exclusive of glycerol.

The difference between the alcohol and the lipids is probably due to the limited number of pathways which would lead to retention of ^{14}C of the labeled ethanol. The greater yield of $^{14}CO_2$ per mMole acyl carbon from labeled unesterified fatty acids in contrast to that derived from the triacylglycerol is probably due to the fact that the former is directly accessible to the fatty acid oxidation enzymes (see Fig. 1). In any case, it is apparent that the intravenously administered triglyceride is metabolized. It is also apparent from the data given in Table 3 that much of the radioactive triolein found in the tissues of the rat three hours following an intravenous injection is subsequently lost during the next 21 hours.

Table 3: Metabolism of intravenously injected $1-^{14}C$ -triolein by the rat.

	Specific Activity of Tissue Lipid*	
	3 Hrs.	24 Hrs.
Liver	4969	239
Kidney	1385	245
Heart	1349	65.4
Lung	2009	507
Spleen	1207	664
Brain	98	55
Muscle	237	67

* C.P.M./Mg lipid.

Presumably a significant amount that disappears is either converted to $^{14}CO_2$ or is transported to the adipose tissue especially from the liver. It is also apparent that the injected lipid has contributed to all of the tissues studied.

A wide variety of triglycerides are used when given intravenously in the form of emulsions (Table 4). All of the substrates listed are converted to other products. As a general rule the shorter the chain length of the constituent fatty acids, the greater the conversion to $^{14}CO_2$. Although medium chain triglycerides might have advantages from this standpoint, to date no commercial intravenous emulsions made with such products are available. Cottonseed and soybean oils are the two most extensively used oils, and of these, only soybean oil is currently most widely employed.

Table 4: Partial list of radioactive substrates used to show parenteral fat is utilized by the rat.

$1\text{-}^{14}C$ Trilinolein	
$1\text{-}^{14}C$ Triolein	All form $^{14}CO_2$,
$1\text{-}^{14}C$ Tripalmitin	^{14}C phospholipids,
$1\text{-}^{14}C$ Trilaurin	and ^{14}C cholesterol
$1\text{-}^{14}C$ Tricaprin	when given intravenously.

When given as unesterified fatty acids, rates of metabolism are considerably faster. Regardless of form given, shorter chain length acids are more completely oxidized to $^{14}CO_2$.

There are many factors other than acyl chain length and size of dose which influence the utilization of intravenous fat emulsions. Some of these are summarized in Table 5. The non-hydrolyzable mineral oil is more slowly removed from the bloodstream than are triglycerides. Similarly, the non-hydrolyzable detergent Triton WR-1339 greatly delays triglyceride removal. In part this effect may be due to substrate level inhibition of lipoprotein lipase. It is no surprise that emulsions which are unstable in the circulation are removed at rates faster than normal; however, this is due mainly to the larger particles being taken out of circulation by being trapped in the small capillaries. As a consequence, the rate of metabolism of the triglyceride is actually decreased. Evisceration and partial hepatectomy decrease the rate at which the fat leaves the blood and also the rate at which it is oxidized. It is of interest to note that while prior fasting in the case of the rat decreases the removal of the fat from the circulation and increases its actual conversion to CO_2, glucose loading has no influence on the disappearance rate, but significantly decreases the oxidation rate. Oral fat loading greatly decreases both the disappearance rate and conversion to CO_2. Administration of heparin accelerates the removal and oxidation rates, while protamine decreases both. The effect of pancreatectomy is of particular interest. In the dog, loss of the pancreas causes almost complete inhibition of emulsified fat disappearance from the blood. Either insulin or heparin can totally overcome this inhibition.

Table 5: Factors affecting parenteral fat utilization in the rat.

	Effect on	
Factor	Removal from Blood	Oxidation to CO_2
Short chain acyl groups	I	I
Mineral oil	D	-
Triton WR 1339	D	D
Large increase in dose	D	D
Repetitive doses	None	None
Instability	I	D
Evisceration	D	D
Hepatectomy	D	D
Fasting	D	I
Glucose load	None	None
Oral fat load	D*	D*
Sex	None	None
Heparin	I	I
Protamin	D	D
+ Pancreatectomy	D	D
+ Pancreatectomy + insulin	Normal	-
+ Pancreatectomy + heparin	Normal	-

I = Increase   D = Decrease
* = Greater for males   + = Dogs

Species differences with respect to the effect of a given variable or removal rate also exist. In contrast to the decrease in rate due to prior fasting noted above with rats, in humans the opposite has been reported to occur (Hallberg, 1965a and Hallberg, 1965b). As seen in Table 6, an almost three-fold increase in the removal rate of Intralipid was found when adults were first fasted 48 hours. It will be noted that the quantity of fat removed under such conditions would furnish approximately 97 calories per kg body weight per 24 hours. For a 70 kg man this would be 6790 per 24 hours! How long such a rapid rate could be maintained or whether the utilization of the fat itself would slow this rate of removal is unknown at present.

Table 6: Rates of removal of Intralipid from bloodstream*.

Age	Hours Fasted	Removal Rate (G. Fat/kg. B.W./24 Hrs)
Pre-Term infants	0	8.6
Adults	overnight	3.8
	38	5.6
	48	10.8

*From Hallberg, 1965a and Hallberg, 1965b.

The question has often been raised as to possible ill effects of the "flooding" of tissues with lipid especially with unesterified fatty acids and/or partial acyl glycerides. From the practical standpoint, it can be pointed out that animals and humans have received very large quantities of Intralipid without adverse reactions. Some of these data are given in Table 7 and are largely compiled from data given by Wretlind (1972). Obviously prudence dictates that moderate quantities of fat be given, especially until much more experience even with Intralipid is accumulated. Two to four grams per kg per 24 hours would appear to be reasonable amounts to administer.

Table 7a: Tolerance of animals to Intralipid.

Species	Dose* (G. Fat/Kg B.W.)	Remarks
Rat**	15 (1)	No deaths or serious effects
	4-8 (14)	No effects
Dog**	15 (1)	No deaths or serious effects
	9 (28)	No effects noted
	9 (70-84)	No effects
Monkey (Rhesus)	4-8 (9)	No adverse effects+

* ( ) Designates number of infusions.
** Transient mild anemia occasionally noted.
+ 10% Intralipid. When 20% Intralipid was used, anemia was found.

Table 7b: Tolerance of Humans to Intralipid.

Age	Dose* (G. Fat/Kg B.W.)	Remarks
Adults	12 (1)	No symptoms noted.
	3 (8-36)	No symptoms noted.
(Crohn's Disease)	2-2.5 (150)	No adverse effects.
(Fulminating Ulcerative Colitis)	12 (15)	No adverse effects.
(Renal Failure)	3-5 (6-31)	No serious effects.
Neonates	2.5-3 (5-11)	Maintained body weight.
(Resections)	3-4 (7-21)	Well Tolerated.
Infants & Neonates	2-4 (?)	Recommended at present.

* ( ) designates number of infusions.

It is unlikely that the body has mechanisms to protect against large or excessive concentrations of unesterified fatty acids and monoglycerides. This contention is reinforced by studies on cells in tissue culture in $vitro$. Cells exposed to the fatty acids or monoglycerides rapidly take up these materials and convert them to droplets of triglyceride if the supply of these substrates exceeds the capacity of the cells to convert them to such products as CO_2 and phospholipids (Lynch and Geyer, 1972). Such accumulation may appear at first glance undesirable. However, by means of its triglyceride synthesizing enzymes the cell is able to take the unesterified fatty acids, which are fairly toxic compounds at all but low concentrations, and rapidly store them as the triesters, in which form they are largely benign (Schneeberger et al., 1971). When the supply of these acids falls, intracellular lipase(s) (Lengle and Geyer, 1972) degrades the triglycerides at rates which supply the cells with non-toxic concentrations of fatty acids and partial glycerides. These products are then used in the normal manner.

These ideas as applied to the infusion of intravenous fat emulsion are summarized in Table 8. Obviously, it remains to be proven that the in $vivo$ counterpart of the mechanisms found in $vitro$ do indeed exist. Certainly the large quantities of intravenously administered fat which can be safely handled by animals and humans suggest that efficient mechanisms to circumvent fatty acid toxicity play an important role.

A large number of different animal species have been used in the tolerance testing of fat emulsions for intravenous use.

Table 8: Intracellular triglyceride formation as a normal deterrent to fatty acid toxicity.

(1) Infused triglycerides are non-toxic as such.

(2) When hydrolyzed to fatty acids, toxicity is averted by:
    (A) Conversion to phospholipids and CO_2.
    (B) Conversion to triglycerides.
        Excess takes form of droplets.

(3) Cellular triglycerides are in constant flux with cellular fatty acids. Composition of droplets changes with time.

(4) If supply of fatty acids decreases, droplets disappear. In such a case the composition of the droplets may change little since the dominant fatty acid available for esterification is derived from droplet.

(5) Intracellular lipolysis is slow enough to avoid fatty acid toxicity.

(6) Intracellular lipid droplets are not primarily for energy storage. Such storage is restricted to adipose cells.

(7) Intracellular lipid droplets serve as a temporary device to avoid fatty acid toxicity.

(8) Intracellular lipid droplets are geared to fatty acid concentration and not to energy supply *per se*.

(9) Intralipid particles which may enter cell as such, would probably undergo controlled lipolysis, thus avoiding toxicity.

These include the rat, mouse, guinea pig, dog, cat and monkey (Geyer, 1960; Wretlind, 1972; Geyer, 1970). Each species has its advantages and disadvantages, and particular aspects of the testing are often best done in certain species. Singularly or *in toto* the results obtained are not necessarily applicable to the human, but this in no way obviates the necessity to carry out the animal tests. The rat and dog have been used most extensively and in the latter species complete parenteral nutrition has been successfully employed. In our studies, the monkey has proven to be the most sensitive species for parenteral emulsion tolerance testing. Of all emulsions examined, only 10% Intralipid was tolerated without any apparent adverse reaction in Macaca mulatta (species of rhesus.) Infusions of 10% Intralipid could be given daily or intermittently with no significant effect on the hematocrit, hemoglobin, or erythrocyte number. With all other emulsions tested a decrease in all of the values was observed, often after just one slow infusion. Simultaneous infusion of amino acids was carried out without incident, provided that crystalline amino acids were used and not protein hydrolysates. Because of the apparent sensitivy of monkeys to intravenous preparations, this species should prove invaluable in the testing of parenteral nutrient solutions and emulsions.

Finally, a unique test of tolerance of the rat for Intralipid has been carried out in our laboratory. We have for some time been engaged in a study directed towards the total replacement of the blood of living animals with an artificial substitute. It was previously reported that such replacement could be achieved, but survival of the rats was only for as long as eight hours (Geyer, 1971). Very recently, however, we have succeeded with modified preparations based on fluorocarbons and polyols not only to completely replace the animal's blood with the artificial preparation, but to have such "bloodless" rats survive, rapidly regenerate blood cells and plasma proteins, and continue to grow and develop. In some of the experiments at the time all normal blood was missing and only the artificial preparation was being circulated, the animals were given parenteral amino acids and Intralipid. No adverse effects of these materials were noted, and the animals subsequently behaved in a normal manner (Table 9). The fact that these nutrient preparations were so well tolerated even though the protective effects of the plasma proteins were lacking, attests to the very low level or absence of toxicity. Further studies on the complete parenteral feeding of "bloodless" animals are planned.

CONCLUSIONS

Fat emulsions for intravenous nutritional purposes have been developed to furnish abundant calories in a small volume without risks inherent in hyperosmotic solutions. They are also a good source of poly-unsaturated acids. A number of different kinds of observations show that such fat is well utilized by a variety of animal species and by humans. Intralipid which is based on soybean oil and egg lecithin has been widely used with few adverse reactions. Even in humans very high doses have not produced ill effects. Of all emulsions tested in monkeys only 10% Intralipid was well tolerated. Simultaneous infusion of solutions of

Table 9: Parenteral Nutrition in "Bloodless" Rats

Experimental Period	Purpose	Description	Remarks
I	Replace all normal blood with an artificial mixture **	Rats totally perfused with an artificial mixture based on fluorocarbon dispersed in polyol-containing aqueous solution	Final hematocrit < 0.1 volume % Final plasma protein < 1.5 mg % Rats continue to live, regenerate cells and proteins, and develop in the usual manner
II	Feed by parenteral means after blood exchange is completed	Animals given: Amino acids * Fat emulsion † Glucose Electrolytes	All nutrients well tolerated. No adverse effects in post-feeding phase

* Aminofusin-L600, J. Pfrimmer and Co., Erlangen, West Germany.

† Intralipid (10%), A. B. Vitrum, Stockholm, Sweden.

** The rats were completely exchange transfused with a protein-free and cell-free synthetic preparation based on high molecular weight polyols and fluorocarbons. All animals survived and subsequently grew and developed in the usual manner.

crystalline amino acids also caused no reactions. Approximately 2-4 grams of fat per kilogram body weight per day appears to be a satisfactory dose for humans at present. Many different factors can influence the rates of utilization and metabolism of intravenously infused fat. For this reason, it is prudent to monitor recipients especially when parenteral feeding extends over many days.

ACKNOWLEDGMENTS

Work concerning the artificial blood substitute studies was supported by a grant from the John A. Hartford Foundation, New York. Intralipid was generously furnished by Dr. Hakansson of Vitrum, Stockholm, Sweden and Aminofusin was generously supplied by Dr. Fekl, Pfrimmer and Co., Erlangen, West Germany.

REFERENCES

Boberg, J. and Carlson, L.A. Clin. Chim. Acta, 10, 420, 1964.

Coats, D.A. and Collins, F.D. Symposium on Intravenous Therapy and Parenteral Nutrition, Melbourne, Australia 1969.

Fraser, R. and Hakansson, I. In: Complete Intravenous Nutrition (A. Wretlind, ed.), S. Karger, Basel, Switzerland, 1972.

Geyer, R.P. Physiol. Rev., 40, 150, 1960.

Geyer, R.P. In: Parenteral Nutrition (H.C. Meng and D.H. Law, eds.), Charles C. Thomas, Springfield, Illinois, 1970.

Geyer, R.P. In: Balanced Nutrition and Therapy (K. Lang, W. Fekl, and G. Berg, eds.), George Thieme, Stuttfart, Germany, 1971.

Hakansson, I., Holm, I., Obel, A.-L., and Wretlind, A. In: Symp. of the Int. Soc. of Parenteral Nutrition (Pallas, Lochham bei Munchen, 1967).

Hallberg, D. Acta Physiol. Scand., 64, 306, 1965a.

Hallberg, D. Acta Physiol. Scand., 65, suppl., p. 254, 1965b.

Hallberg, D. In: Parenteral Nutrition (H.C. Meng and D.H. Law, eds.), Charles C. Thomas, Springfield, Illinois, 1970.

LeKim, D., Betsing, H. and Stoffel, W. Hoppe-Seyler's Z. Physiol. Chem. Bd., 353, 949, 1972.

Lengle, E. and Geyer, R.P. Biochim. Biophys. Acta, 260, 608, 1972.

Lynch, R.D. and Geyer, R.P. Biochim. Biophys. Acta, 260, 547, 1972.

Schneeberger, E.E., Lynch, R.D. and Geyer, R.P. Exptl. Cell. Res., 69, 193, 1971.

Schoefl, G.I. Proc. Roy. Soc. B., 169, 147, 1968.

Wretlind, A. In: Complete Intravenous Nutrition (A. Wretlind, ed.), S. Karger, Basel, Switzerland, 1972.

ALCOHOL METABOLISM DURING DEVELOPMENT

Esteban Mezey, M.D.

The Alcoholism Research Unit & Department of Medicine,
Baltimore City Hospitals and The Johns Hopkins University
School of Medicine, Baltimore, Maryland 21224

Ethanol has been used recently in obstetrics and in neonatal pediatrics for three principal therapeutic reasons: prevention of premature labor, reduction of anticipated neonatal hyperbilirubinemia, and parenteral nutrition. The use of ethanol in the prevention of premature labor is based on its inhibitory effect on uterine activity in early labor; it presumably acts by inhibiting the release of oxytocin from the neurohypophysis (Fuchs et al., 1967). Ethanol when administered to near-term or term pregnant patients has been shown to reduce hyperbilirubinemia in the neonates (Waltman et al., 1969). Its use has been suggested, therefore, to reduce anticipated hyperbilirubinemia in those neonates in which high serum bilirubin may be expected, such as premature infants. The reduction of bilirubin after ethanol administration is probably due to increased glucuronidation of bilirubin since ethanol has been demonstrated to stimulate a variety of microsomal enzymes in both animals and man (Rubin et al., 1968). Finally, ethanol has been used extensively for parenteral nutrition in premature neonates (Babson, 1971), as well as in adult surgical patients (Karp et al., 1951). Ethanol when oxidized provides seven calories per gram. Its addition to parenteral solutions containing glucose and protein hydrolysate allows for a significant caloric content of the solution with little change in its volume. Babson (1971), for instance, has reported feeding low birth weight infants with parenteral solutions containing 2% protein hydrolysate, 10-13% glucose and 1.5% ethanol which provide 65 calories per 100 ml and are given at flow rates of 12 ml per hour for one to four days. Ethanol in this solution contributes 16% of the total calories administered.

Because of the uses of ethanol in the newborn and in pregnant patients, where it readily transverses the placental barrier and becomes distributed evenly in the maternal and fetal blood stream (Belinkoff and Hall, 1950), it seems important that the capabilities of ethanol elimination by both fetus and newborn be known.

Ethanol, after its administration, is distributed rapidly and evenly throughout the whole body water and appears in the urine, pulmonary alveolar air and spinal fluid. Its elimination from the body is accomplished chiefly by its metabolism mainly in the liver; less than 10% of an ingested dose is lost through the kidney, lungs, or skin The first step in the metabolism of ethanol is its oxidation to acetaldehyde, which is rate limiting. The principal enzyme that catalyzes this oxidation is alcohol dehydrogenase present in the supernatant fraction of liver homogenates (Nyberg et al., 1953). Recently, the oxidation of ethanol by a microsomal enzyme system was also demonstrated (Orme-Johnson and Ziegler, 1965; Lieber and DeCarli, 1968). However, some investigators have suggested that the oxidation of ethanol observed in microsomes is due to contaminating catalase in the presence of hydrogen peroxide generated by NADPH oxidase and oxygen (Isselbacher and Carter, 1970; and Thurman et al., 1972).

Studies of rates of ethanol metabolism and ethanol oxidizing enzymes in the fetus and newborn have been few (Table I).

Table 1: Ethanol Metabolism During Development

Investigators	Subjects	Ethanol Metabolism	
		mg/100 ml/hr	mg/kg/hr
Pikkarainen (1971)	Fetal Livers (10-12 weeks) Liver perfusion	0	-
Wagner (1970)	5 premature infants (2000-2500 g)	7.4	77
	1 full-term infant (low wt. of 2000 g)	11.0	110
Seppälä (1971)	2 premature twins (1100 g and 1940 g)	8.0 & 7.0	-
	Mother	14.0	
	Normal Adults	15.0	100

Pikkarainen (1971) was unable to demonstrate any ethanol oxidation
by livers of 10-16 week old human fetuses. Perfusion of the livers
with a medium containing ethanol at concentrations of 0.5 mM and 1.0
mM did not result in the formation of detectable acetaldehyde or
change in lactate/pyruvate ratio. Wagner et al. (1970) determined
the rates of ethanol disappearance from the blood in six healthy
premature infants weighing between 2000 and 25000 g 23-53 hours
after delivery. The ethanol was administered intravenously in a
dose of 0.4 to 0.95 g per kg as a 10% solution. In five premature
infants the rates of ethanol disappearance from the blood averaged
7.4 mg per 100 ml per hour corresponding to 77 mg per kg body
weight per hour, while in the sixth infant who weighed 2000 g but
was of 40 weeks gestation, the rate was 11 mg per 100 ml per hour
or 110 mg per kg body weight per hour. The mean rate in the premature newborns is about half the average value of 15 mg per 100 ml
per hour or 100 mg per kg body weight per hour found in normal
adults. In another report by Seppälä et al. (1971), rates of
ethanol disappearance were determined in premature twins born at
32 weeks gestation after the mother had received ethanol in an
attempt to reduce uterine contractions. The weights of the newborns were 1100 g and 1940 g and their rates of ethanol disappearance
from the blood, 8 and 7 mg per 100 ml per hour, respectively,
compared with a rate of 14 mg per 100 ml per hour in the mother.

The activity of alcohol dehydrogenase at various stages of
development of the human liver was determined by Pikkarainen and
Räihä (1967). Alcohol dehydrogenase activity in the livers of
two-month old fetuses was only 3-4% of that found in adult livers.
There was a gradual increase in activity during fetal development
and at birth the activity had increased to 10-15% of the adult.
However, adult activities were not reached until after five years
of age. The alcohol dehydrogenase activity in the fetal liver
was shown to differ from that in the adult liver in having a pH
optimum for ethanol oxidation of 10.0 compared with 10.4, and
higher Michaelis-Menten constants for ethanol and NAD^+. Also,
while four distinct isoenzyme bands could be obtained on starch
gel electrophoresis of the adult liver, only one was present in
the two-month old fetal liver and two in the full-term newborn
(Pikkarainen and Räithä, 1969).

Recently, determinations of alcohol dehydrogenase, microsomal
ethanol oxidizing activity, catalase, cytochrome P-450, and
aniline hydroxylase were performed in our laboratory on fetal
livers at various stages of development. The fetuses were obtained
by hysterectomy from legal abortions. The livers were immediately
removed, weighed, washed in ice-cold physiologic saline, homogenized,
and the supernatant fractions and washed microsomal pellets prepared as described previously (Tobon and Mezey, 1971). The enzyme
activities were determined as follows: Catalase in the homogenate

(Feinstein, 1949), alcohol dehydrogenase in the supernatant fraction (Bonnichsen and Brink, 1955), and the NADPH dependent ethanol oxidizing activity (Lieber and DeCarli, 1968), cytochrome P-450 (Omura and Sato, 1964), and aniline hydroxylase (Imai and Sato, 1966) in the microsomal fraction The ages of the fetuses were estimated according to Scammon and Calkins (1923). Alcohol dehydrogenase (Table II) showed a gradual increase in activity throughout development from 0.10 to 0.57 μmoles per mg protein per hour in fetuses from 10 to 20 weeks in age,

Table II: Fetal Development of Ethanol Oxidizing Enzymes

Fetal Age	Alcohol Dehydrogenase	Microsomal Ethanol Oxidizing Enzyme	Catalase
wks	μmoles/mg*/hr	nmoles/mg/min	mEq/mg/5 min
10	0.10	11.0+	0.41
13	0.29	4.9	0.25
17	0.24	5.6	0.55
19	0.29	3.3	0.34
20	0.37	3.5	0.35
20	0.57	4.4	0.50

*Enzyme activities are expressed per mg of protein.

+The microsomal pellet was very small and was not washed, resuspended and recentrifuged prior to assay.

These values compare with a mean adult activity of 2.1 ± 0.3 (SE) μmoles per mg per protein per hour (Mezey et al., 1968). Both the microsomal ethanol oxidizing enzyme system and catalase activities were detected as early as 10 weeks of fetal age, and by contrast to alcohol dehydrogenase, no increases in activity were found during development. Cytochrome P-450 was detected at 13 weeks (Table III), and reached adult levels (Alvares et al., 1969) at 17 weeks of age. The activities of aniline hydroxylase obtained in four fetal livers were comparable to those reported in adult livers (Darby et al., 1970), The significant activities of the hepatic microsomal drug metabolizing enzymes found in early fetal development by us and other investigators (Yaffe et al., 1970) are in sharp contrast to the low or absent activities found in several animal species during fetal and newborn development (Jondorf et al, 1958).

Table III: Fetal Development of Cytochrome P-450 and Aniline Hydroxylase Activity.

Fetal Age	Cytochrome P-450	Aniline Hydroxylase
wks	μmoles/mg*	nmoles/mg/min
10	0	-
13	0.06	9.8
17	0.40	-
19	0.25	2.6
20	0.43	3.9
20	-	2.6

*Values expressed per mg of protein.

In conlcusion, the rates of ethanol metabolism and the activity of hepatic alcohol dehydrogenase are both reduced in fetal life and in the newborn when compared with values obtained in the adult. Significant microsomal ethanol oxidizing system and catalase activities are detected early in fetal life and do not increase during further fetal development However, the contribution of these two enzymes to the in vivo metabolism of ethanol during development remains to be evaluated.

REFERENCES

Alvares, A.P., Schilling, G., Levin, W., Kintzman, R., Brand, L., and Mark, L.C. Clin. Pharmacol. Ther., 10, 655, 1969.

Babson, S.G. J Ped., 79, 694, 1971.

Belinkoff, S. and Hall, O.W., Jr. J. Obstet. Gynec., 59, 429, 1950.

Bonnichsen, R.K. and Brink, N.G. in Methods in Enzymology, Vol. I, (ed. S.P. Colowick and N.O. Kaplan), New York, 1955, Academic Press, p. 496.

Darby, F.J., Newnes, W. and Prince Evans, D.A. Biochem. Pharmacol., 19, 1514, 1970.

Feinstein, R.N. J. Biol. Chem., 180, 1197, 1949.

Fuchs, F., Fuchs, A.-R., Poblete, V.F., Jr. and Risk, A. Am. J. Obstet. Gynec., 99, 627, 1967.

Imai, Y. and Sato, R. Biochem. Biophys. Res. Comm., 25, 80, 1966.

Isselbacher, K.J. and Carter, E.A. Biochem. Biophys. Res. Comm., 39, 530, 1970.

Jondorf, W.R., Maickel, R.P. and Brodie, B.B. Biochem. Pharmacol., 1, 352, 1958.

Karp, M. and Sokol, J.K. JAMA, 146, 21, 1951.

Lieber, C.S. and DeCarli, L.M. Science, 162, 197, 1968.

Mezey, E., Cherrick, G.R. and Holt, P.R. New Eng. J. Med., 279, 241, 1968.

Nyberg, A., Schuberth, J. and Anggard, L. Acta Chem. Scand., 7, 1170, 1953.

Omura, T. and Sato, R. J. Biol. Chem., 239, 2370, 1964.

Orme-Johnson, W.H. and Ziegler, D.M. Biochem. Biophys. Res. Comm., 21, 78, 1965.

Pikkarainen, P.H. and Räihä, N.C.R., Ped. Res., 1, 165, 1967.

Pikkarainen, P.H. and Räihä, N.C.R, Nature, 22, 563, 1969.

Pikkarainen, P.H. Life Sci., 10, 1359, 1971.

Rubin, E. and Lieber, C.S. Science, 162, 690, 1968.

Scammon, R.E. and Calkins, L.A. Proc. Soc. Exp. Biol Med., 20, 353, 1923.

Seppälä, M., Räihä, N.C.R. and Tamminen, V., Lancet, 1, 1188, 1971.

Tobon, F. and Mezey, E.   J. Lab Clin. Med., 77, 110, 1971

Thurman, R.G., Ley, H.G. and Scholtz, R.   Eur. J. Biochem., 25, 420, 1972.

Waltman, R., Nigrin, G. and Pipat, C.   Lancet II, 1265, 1969.

Wagner, L., Wagner, G. and Guerrero, J.   Am. J. Obstet. Gynec., 108, 308, 1970.

Yaffe, S.H., Rane, A., Sjoqvist, F., Boreus, L. O. and Orrenius, S.   Life Sci., 9, 1189, 1970.

FLUID AND ELECTROLYTE REQUIREMENTS AND TOLERANCE

John D. Crawford

Children's Service, Massachusetts General Hospital

Boston, Massachusetts 02114

Twenty-five years association with the Children's Service, much of it concerned with how to devise better means to support the child during a period when critical illness precludes his taking nourishment in the ordinary fashion, have taught me to approach the assigned topic with caution and humility. One lesson learned is that neither requirement nor tolerance for any individual component of a parenteral fluid mixture is constant, a change in one affecting the metabolism of others. Perhaps even more obviously, the results of experiments on normal adult individuals cannot be directly extrapolated to the sick infant for age, size, medications and the illness itself affect both requirement and tolerance. Because my own particular focus has been principally on water metabolism, the time allotted will be devoted to a personal review of the development of knowledge in this circumscribed area attempting to make clear the need for a dynamic approach to the handling of any particular component of a parenteral fluid mixture, be it water, glucose, fat or phosphate.

When I joined the Department of Pediatrics at the Massachusetts General Hospital, Drs. Butler, Talbot and Gamble were at work devising their life-raft ration for castaways (Gamble and Butler, 1944). In the course of these studies it was of importance to define the minimal water requirement for urine formation. Their elucidation of this parameter prompted Dr. Edgar Schoen, and myself (Crawford et al., 1952) greatly aided by Dr. Arnold Nicosia's introduction of the Thermistor for osmometry (Crawford and Nicosia, 1952) to look at the other side of the coin, to see if there existed a maximal renal capacity for water excretion. As shown in Figure 1, this too, proved to be a finite quantity. When normal subjects and patients were given increasing amounts of water, a sharp point was reached where urine flow and dilution failed to increase and additional water caused hemodilution and weight gain.

The demonstration of these two parameters, the renal water requirement and maximal excretory capacity (Figure 2) led to the concept of "floors" and "ceilings" which Dr. Talbot has so

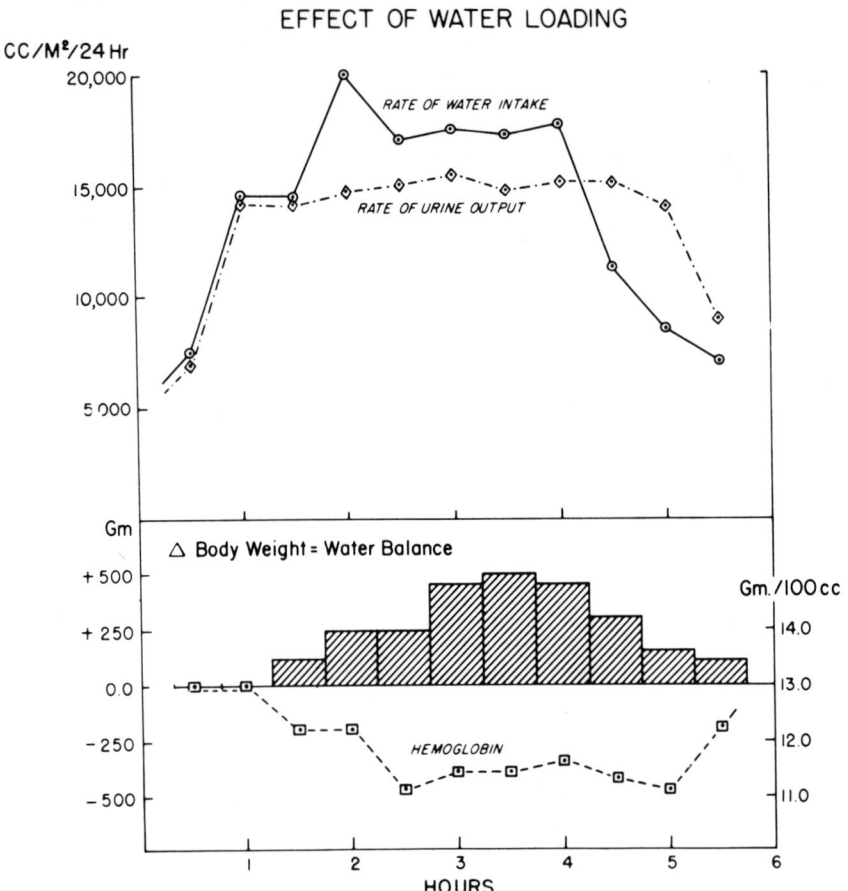

Fig. 1. Demonstration of maximum tolerance for water. The water intake of this lad was increased stepwise. At a rate equivalent to a daily intake of 15 liters, urine flow ceased to rise in parallel and a positive water balance and dilutional anemia developed. Because the patient had diabetes insipidus, he failed to exhibit the adaptive delay shown by the subject of Fig. 7. (From Talbot, N.B., Sobel, E.H., McArthur, J.W. and Crawford, J.D. <u>Functional Endocrinology from Birth to Adolescence</u>, Harvard University Press, Cambridge, Massachusetts, 1952).

successfully employed in his writings (Talbot et al., 1953; Kerrigan et al., 1955; Talbot et al., 1956; Talbot et al., 1957; Talbot et al., 1959). It quickly became evident, however, that these limits were importantly modified by illness.

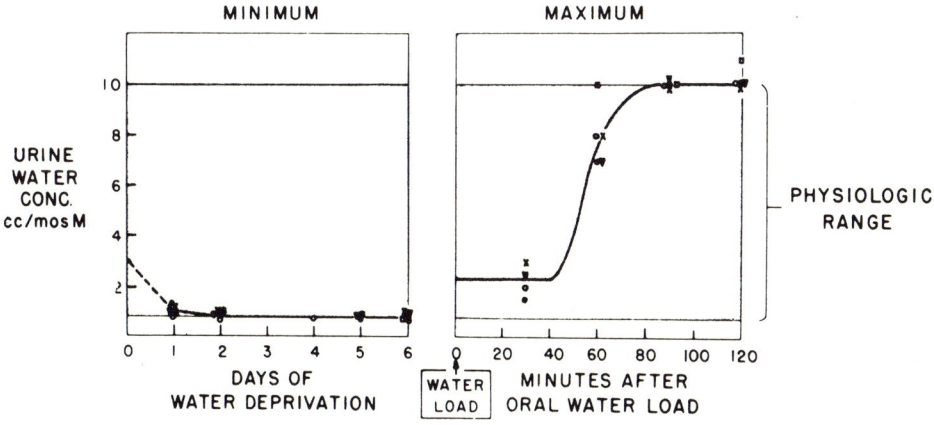

Fig. 2. Normal limits of renal water conservation and excretion. The data shown on the left are from the water deprived castaways of Gamble and Butler (1944); those on the right are from normal individuals loaded with water much as was the subject of Fig. 1. The ordinate scale indicates "water concentration", or milliliters of water accompanying each solute millosmole in urine (Kerrigan et al., 1955).

Figure 3 taken from his syllabus, Metabolic Homeostasis (Talbot et al., 1959) depicts observations on a patient with advanced renal insufficiency. The kidney ailment compromised his ability either to concentrate or to dilute his urine with the result that his "ceiling" of water tolerance was reduced to about one-fifth of normal and his "floor" was elevated to twice that pertaining in the absence of kidney disease.

Other non-renal factors modifying the water requirement and tolerance of the ill patient must spring to the clinicians mind if his prescription of parenteral fluids is to maintain the patient in the safe zone where the automatic mechanisms for retention and

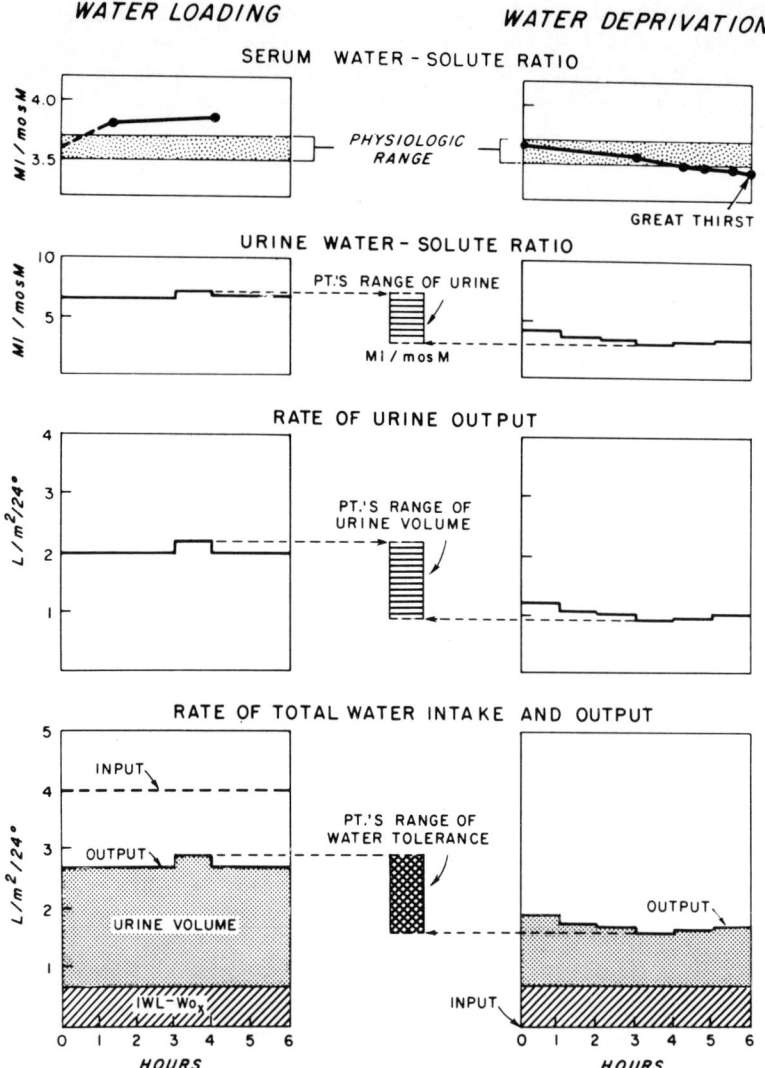

Fig. 3. Alteration by kidney disease of water requirement and tolerance. The patient whose studies are represented here was loaded with water (left) and deprived (right) until the water concentration of serum (top) increased or fell below normal limits (shaded zone). The corresponding urine water concentrations (second set of panels) define the limits of adjustment. Maximal and minimal urine flows are shown in the third set of panels from the top. The patient's maximum water tolerance and minimal requirements are given in the bottom panels which integrate the insensible loss and water gained from metabolic oxidations (Talbot et al., 1959).

disposal can preserve homeostasis. The increments in water requirement of fever or of the Kussmaul respiration of the diabetic in ketoacidosis hardly need mention. Drs. Schoen, Nicosia and I (1952) in working out the factors influencing the maximal capacity for water excretion in urine "rediscovered" a phenomenon earlier described by Professor McCance and Miss Widdowson (1937); we noted that urinary total osmotically active solute excretion was very low, while urinary dilution was normally efficient, the maximal volume of water eliminated was much reduced. As Figure 4 shows, this is because it is a direct function of the rate of osmolar excretion.

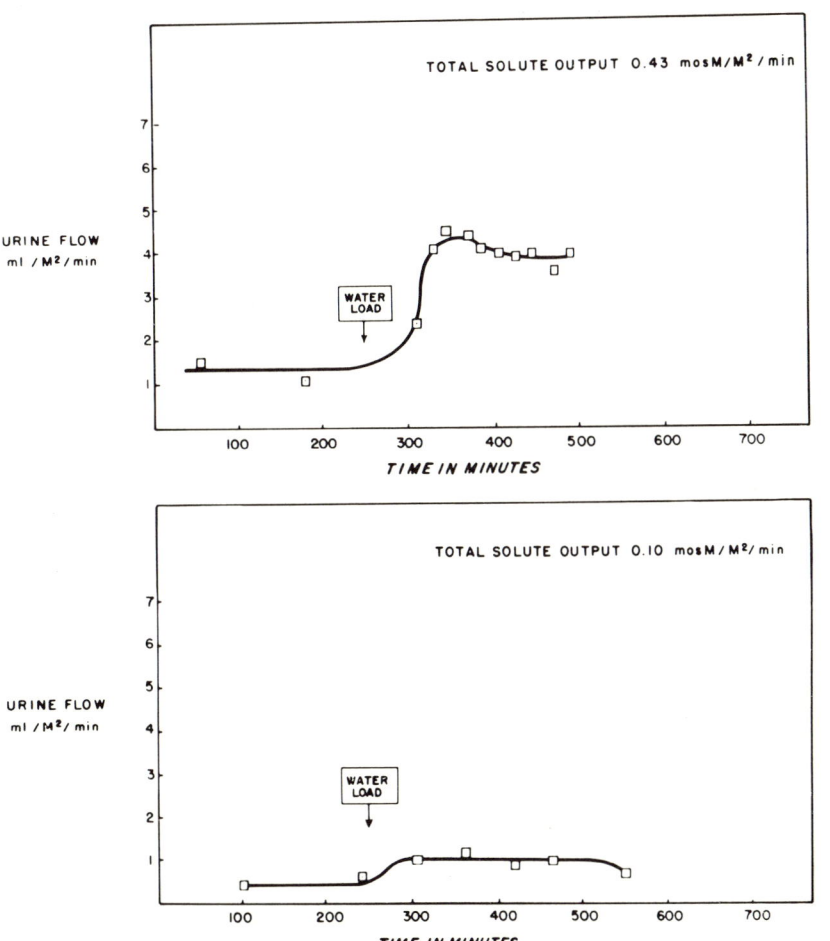

Fig. 4. Influence of solute output on urine flow. The subject of this study had a normal ability to dilute his urine in response to water loading. The restricted maximal rates of urine formation were a function of special low solute residue diets (Crawford et al., 1952).

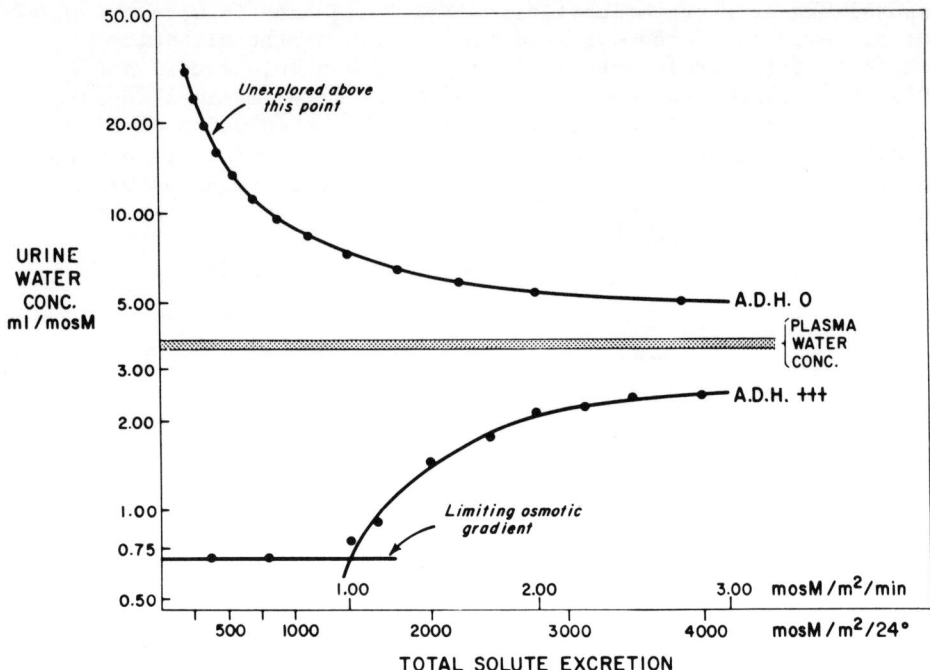

Fig. 5. Alterations in maximal urine dilution and concentration with changes in solute excretion. The upper curve (A.D.H. 0) is derived from observations on water loaded subjects. The lower (A.D.H. +++) from data obtained during hydropenia. The wide variation in solute excretion resulted from dietary preparation and intravenous saline or mannitol loading (Crawford et al., 1957).

On the other hand, at high rates of solute extraction, both the dilution and concentration maxima were reduced (Figure 5), each of the two parameters appearing to approach 4 ml per mosM (Crawford et al., 1957). This, of course, is the concentration of preformed urine as it reaches the distal convoluted tubule for the final tonicity adjustments under the influence of vasopressin.

The studies leading to the development of the curves of Figure 5 involved leading with a variety of different solutes, principally sodium chloride, mannitol and urea. We noticed that the latter stood out as different in the sense that a greater degree of water economy was achieved when urea was the predominant osmotically active urinary solute. Again, when Dr. Jan Probst and I addressed ourselves to the mechanism underlying this phenomenon we realized we had made another rediscovery and could find no better title for our report in 1959 (Crawford et al., 1959) than the one Dr. Gamble had used in 1934, "The Service of Urea in Renal Water Conservation" (Gamble et al., 1934).

Medicine seems never to allow for complacency. As illustrated in Figure 6, we have been trapped in the Charybodian vortex through failure to recognize that the unusually safe rates of parenteral fluid administration may become quite inappropriate for the surgical patient as a result of operation itself, anesthesia, and the use of opiates for pain.

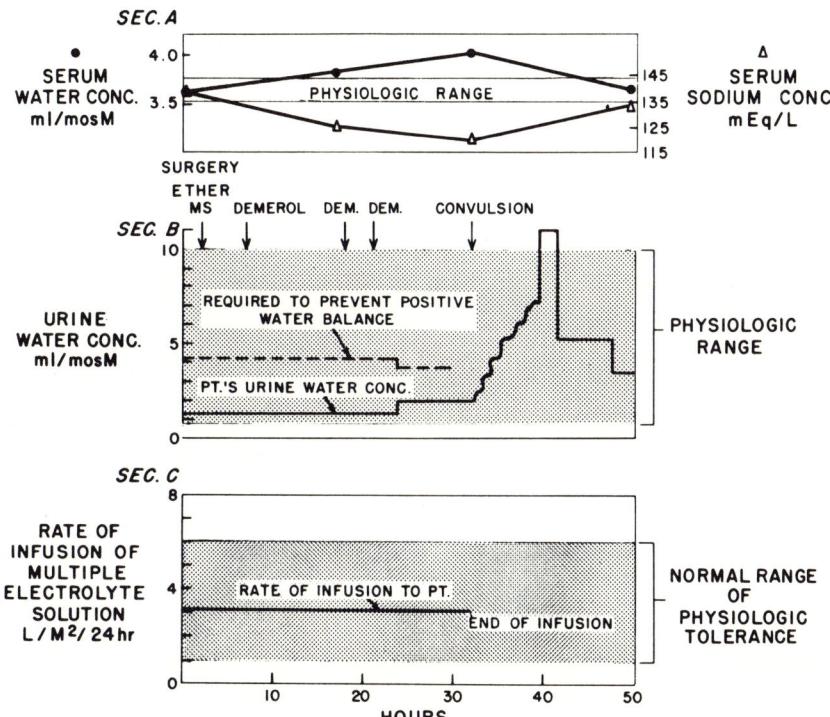

Fig. 6. Seizures resulting from an inappropriately high rate of intravenous water delivery in a child undergoing appendectomy. Until the time when seizures developed, the patient received parenterally a multiple electrolyte fluid at the rate of approximately 3 liters per m^2 per 24 hrs., those in charge failing to recognize the special influences of operation, ether anesthesia, morphine and demerol to depress the diuretic response. Hyponatremia and an increase in serum water concentration developed ultimately leading to water intoxication and seizures. (Kerrigan et al., 1955).

The experience of inducing seizures due to water intoxication in the little girl of Figure 6 undergoing a routine appendectomy led to a collaborative project with Dr. Bunker and his group in anesthesia to better define the water tolerance of patients undergoing an operation (Aprahamian et al., 1959) as well as to a series of studies with Drs. Philip Dodge, Juan Sotos and Gwendolyn Hogan on the impact on brain function of hyper- and hypotonic states (Crawford and Dodge, 1959; Dodge et al., 1960; Sotos et al., 1960; Hogan et al., 1972).

Yet another factor which must enter consideration as to what constitutes safe parenteral fluid and electrolyte administration is adaptation time. Figure 7 shows the accumulation and disposal of water in a small child in whom the rate of delivery of intravenous fluids was increased abruptly to 30 ml per minute for half an hour, then slowed to 10 ml per minute for 90 minutes before being reduced to a bare maintenance level.

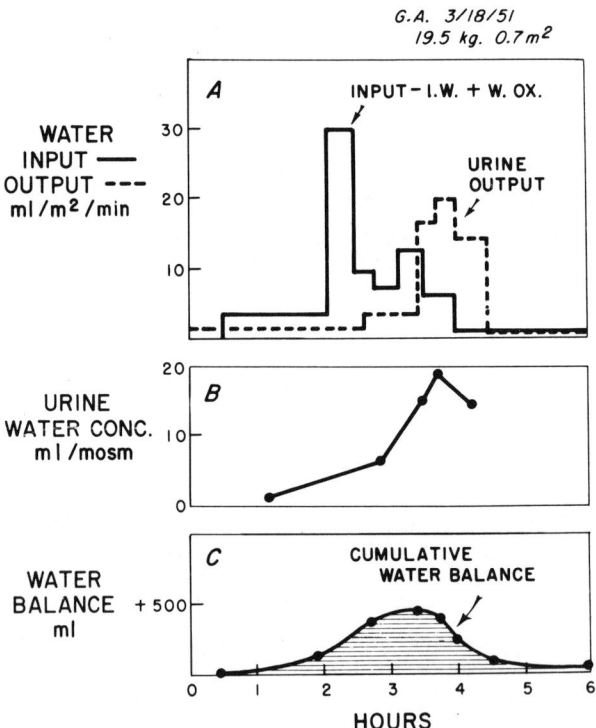

Fig. 7. Dynamics of adaptation to a change in rate of water delivery. These data were obtained on a child with minimal lesion nephrotic syndrome at a time when his serum total protein concentration was not so low as to significantly increase the adaptation time to water loading (Talbot et al., 1957).

The delay in the adaptive response of about an hour's time is readily apparent. This system which involves dilution of body fluids sensed by the hypothalamic osmoreceptors which in turn shut off vasopressin release is one of the body's most nimble. Figure 8 contrasts the much more sluggish response of the system concerned with adapting to a sudden change in sodium intake where achievement of a new equilibrium requires five days.

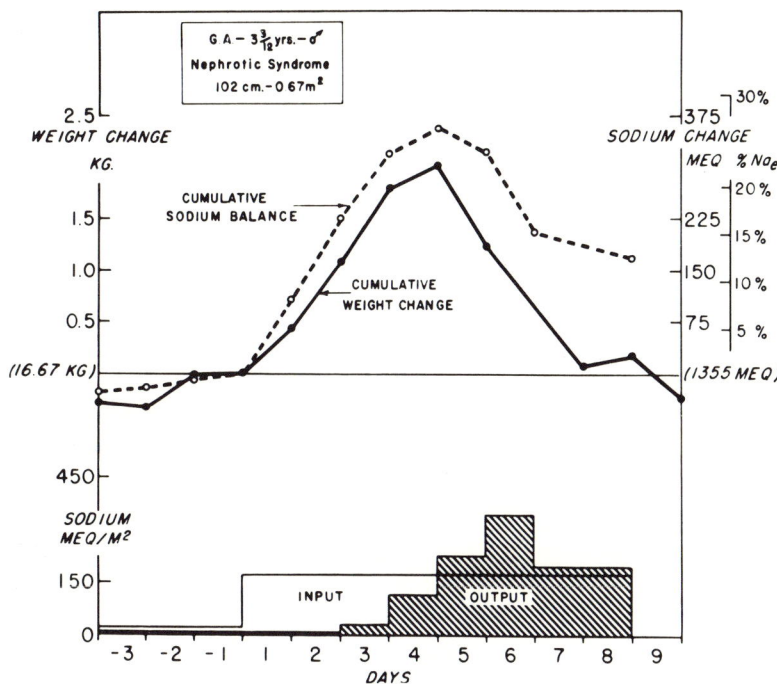

Fig. 8. Dynamics of adaptation to a change in sodium intake. Observations on the same child with nephrosis as depicted in Fig. 7. Loading was with oral sodium chloride with adaptation complete on the 7th day.

As the importance of these adaptations began to be appreciated, Dr. Neyzi undertook a close examination of the convenient hospital practice of providing patients their full daily fluid and electrolyte requirements by means of brief intravenous infusions and adduced impressive evidence that this practice cause major dislocations of body composition (Neyzi et al., 1958). The recent introduction by Drs. Wilmore and Dudrick (1968) of "hyperalimentation" using parenteral glucose and protein hydrolysates to meet the full caloric requirement has prompted all of us to be mindful of the time required for adaptation to altered carbohydrate administration. Neglect of the dynamics of this regulatory system leads to osmotic diuresis at the start of full parenteral alimentation while abrupt discontinuation risks hypoglycemic seizures.

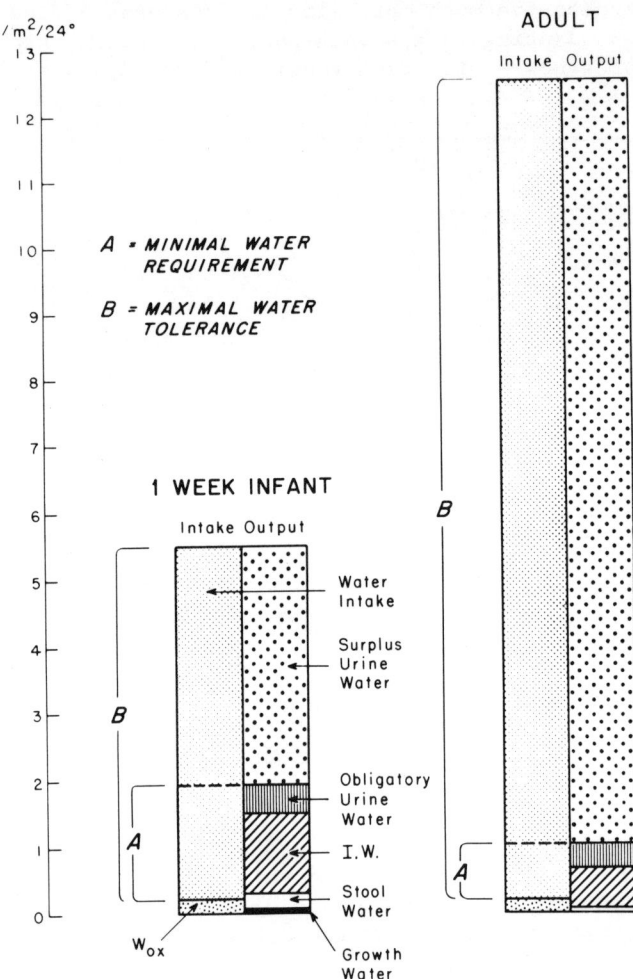

Fig. 9. The water requirement and tolerance of the infant as compared with those of the adult. Note that the infant's minimal water requirement is increased by relatively larger insensible and stool losses and an appreciable diversion of water to new protoplasm for growth. The water tolerance of the neonate is reduced by a slight reduction in the renal capacity to dilute accompanied by a substantially prolonged time for adaptation to a water load. Tolerance is also variably reduced by diet; breast milk, for example, limits water excretion capacity because of its very small osmotically active residue. (Talbot et al., 1959).

Figure 9 introduces one last consideration, that of the body size. As Dr. Gamble (1958) emphasized, the body reservoirs of water, electrolytes and calories are large relative to turnover in the adult. In the small infant, however, daily water turnover approximates 20 per cent of the body water pool. Thus, only one day's unreplaced losses may culminate in dehydration of the greatest degree compatible with life. Coupled with this are the considerably larger minimal daily requirements and lesser tolerances of a variety of essential nutrients when, as in Figure 9, these are mathematically normalized per unit of body surface area. Accordingly, it is perhaps little to be wondered at that those concerned primarily with the care of infants and children have had pressed upon them a leadership role in defining the limits of safety in provision of parenteral fluids and electrolytes and to develop means of full caloric support through periods of critical illness when the usual functions of the digestive tract are compromised.

REFERENCES

Aprahamian, H.A., Vanderveen, J.L., Bunker, J.P., Murphy, A.J. and Crawford, J.D. Ann. of Surg., 150, 122, 1959.

Crawford, J.D., Schoen, E. and Nicosia, A.P. Amer. J. Dis. Child., 83, 91, 1952.

Crawford, J.D. and Nicosia, A.P. J. Lab. and Clin. Med., 40, 907, 1952.

Crawford, J.D., Cushman, A.M., Parisi, A. and Terry, M.L. J. Clin. Invest., 36, 880, 1957.

Crawford, J.D., Doyle, A.P. and Probst, J.H. Amer. J. Physiol., 196, 545, 1959.

Crawford, J.D. and Dodge, P.R. Ped. Clin. N. Amer., 6, 257, 1959.

Dodge, P.R., Crawford, J.D. and Probst, J.H. Arch. of Neurology, 3, 513, 1960.

Gamble, J.L., McKhann, C.F., Butler, A.M. and Tuthill, E. Amer. J. Physiol., 109, 139, 1934.

Gamble, J.L. and Butler, A.M. Trans. Assn. Amer. Phys., 58, 157, 1944.

Gamble. J.L. Chemical Anatomy, Physiology and Pathology of Extracellular Fluid; A Lecture Syllabus (6th Ed.), Harvard University Press, Cambridge, Massachusetts, 1958.

Hogan, G.R., Dodge, P.R., Gill, S.R., Scholl, M.L. and Master, S. Pediatrics, 50, 769, 1972.

Kerrigan, G.A., Talbot, N.B. and Crawford, J.D. J. Clin. Endocr., 15, 265, 1955.

McCance, R.A. and Widdowson, E.M. J. Physiol., 91, 222, 1937.

Neyzi, O., Bailey, M. and Talbot, N.B. New Eng. J. Med., 258, 1239, 1958.

Sotos, J.F., Dodge, P.R., Meara, P. and Talbot, N.B. Pediatrics, 26, 925, 1960.

Talbot, N.B., Crawford, J.D. and Butler, A.M. New Eng. J. Med., 248, 1100, 1953.

Talbot, N.B., Crawford, J.D., Kerrigan, G.A., Hillman, D.A., Bertucio, M.L. and Terry, M.L. New Eng. J. Med., 255, 655, 1956.

Talbot, N.B., Crawford, J.D. and Cook, C.D. New Eng. J. Med., 256, 1080, 1957.

Talbot, N.B., Richie, R.H. and Crawford, J.D. Metabolic Homeostasis, A syllabus for those concerned with the care of patients. Harvard University Press, Cambridge, Mass. 1959.

Wilmore, D.W. and Dudrick, S.J. J.A.M.A., 203, 860, 1968.

TRACE ELEMENTS AND VITAMINS

Harry L. Greene, Michael Hambidge, and Yaye F. Herman

Metabolic Division, U.S. Army Medical Research and
Nutrition Laboratory, Fitzsimmons General Hospital
Denver, Colorado 80240

What might appear to be nutritionally unimportant or of only minor importance in the adult can be extremely important in the infant who may increase his body mass by 50 to 75 per cent during a few weeks of total intravenous nutrition. This dilutional factor alone might be enough to significantly deplete body stores of some micronutrients.

During this period of rapid somatic growth, there is a high rate of brain growth and evidence further suggests that certain vitamins such as folate and riboflavin and at least one trace element, zinc, may be particularly active in the normal process of protein synthesis and cell replication.

For these reasons, it seems imperative that if one is to have a solution for complete intravenous nutrition in the infant, studies must be done to determine the requirements of both the macro- and micronutrients.

TRACE ELEMENTS

At this time there is evidence that in addition to the eleven major elements, thirteen trace elements are nutritional requirements for mammals and/or birds. These are: iron, copper, zinc, vanadium, chromium, nickel, manganese, molybdenum, cobalt, tin, selenium, silicon and iodine. Additional elements are almost certain to be added to this list of essential micronutrients within the next few years. Information on some of these trace elements is summarized in Table 1. Listed are zinc, copper, chromium, iron and iodine because thus far, deficiencies of these elements are already known to cause symptoms in man under certain circumstances (Sandstead et al., 1972).

Table 1: Trace elements essential in human nutrition

	Oral Requirements	Intravenous Requirements*	Deficiency Symptoms
Zinc	12 mg/day (adults)	10-30 µg/kg	Anorexia, growth retardation, impaired reproductive development, poor wound healing, keratogenesis, hypogeusia.
Copper	42-135 µg/kg/day	15-22 µg/kg	Anemia, leukopenia, bone abnormalities.
Chromium	100 µg/day** (adults)	0.2 µg/kg	Abnormal glucose tolerance.
Iodine	5 µg/kg	4-5 µg/kg	Hypothyroid goiter.
Iron	10-20 mg/day	1 mg/day	Anemia, weakness.

* Estimated Requirements - Unconfirmed.

** Total dietary Chromium, uncertain amount of which is "glucose tolerance factor" chromium.

(From Underwood, 1972).

With the possible exception of iron and iodine, daily oral requirements for adults have not been accurately defined. Corresponding requirements for the infants and young growing child are generally even less certain. There is evidence from animal studies, that the young subject's requirements for some of these elements, e.g., zinc, are relatively high. The figures shown for oral requirements are, therefore, only approximate. The percentage of absorption for many of these elements is certainly less than 50% and is quite variable. It depends on many factors, e.g., the quantity in the diet, overall balance, interference from other trace elements, phytate, calcium, and phosphate. If the minimal daily absorption required to maintain balance were known for the trace elements as it is for iron, e.g. approximately 1 mg of iron per day for an adult man, an approximate figure might be arrived at for the intravenous requirements during total parenteral nutrition. However, possible differences resulting from different routes of absorption and different chemical forms of the elements have to be considered. Thus, there is little accurate information available at present on intravenous requirements for these elements. The figures given in this table are similar to those which have been used by Shils (1972) with no apparent adverse sequelae.

Zinc deficiency has been described in many species, including man; it is characterized by anorexia, growth retardation, delayed sexual maturation, poor wound healing and ketatogenesis. Recent evidence indicates that it is also associated with impaired taste acuity in man (Hambidge et al., 1972). Severe multiple congenital malformations occur in the offspring of zinc deficient maternal rats. Apart from a requirement for at least 20 zinc metalloenzymes this element is also required for nucleic acid metabolism and protein synthesis, e.g. the incorporation of thymidine into DNA is suppressed by zinc deficiency with a concommitant decreased nitrate activity in the neural crest of laboratory animals (Swenerton et al., 1969). This observation further emphasizes the importance of zinc during critical periods of growth.

Several recent reports document a symptomatic copper deficiency in premature infants and have included one subject receiving total parenteral alimentation. Premature infants have relatively small hepatic copper stores and probably require more than the term infant (Cordano et al., 1964). Copper deficiency is generally seen in association with iron deficiency, but the anemia and weakness do not respond to iron therapy alone.

Trivalent chromium, which acts as a co-factor for insulin at the cell membrane, is required for normal glucose tolerance. In clinical trials a number of patients with maturity onset diabetes and elderly patients with impaired glucose tolerance have had significant improvement in glucose tolerance following oral chromium ($CrCl_3 6H_2O$) supplementation of 150-1,000 micrograms per day for several weeks or months (Levin et al., 1968). Requirements for chromium can depend on the chemical form of this element. The biological activity of inorganic trivalent chromium and the percentage absorption (approximately 1%) is very much less than that of chromium organically bound in the "glucose tolerance factor" coumpound. The structure of the latter awaits complete elucidation and purified preparations are not yet available for human administration. It is possible that the human infant may utilize inorganic chromium more effectively than the adult. This is suggested by the rapid improvement in glucose tolerance of children with protein-calorie malnutrition in Jordan, Nigeria and Turkey, following a single oral supplement of 250 µg inorganic trivalent chromium

(Hambidge, 1972). A glucose load leads to a rise in plasma chromium, and much of this chromium may then be lost in the urine. It is possible, therefore, that the large amounts of glucose infused during total parenteral alimentation may result in an increased requirement for chromium.

Iodine requirements in the infant are not precisely known but it is efficiently absorbed and therefore the oral recommended dose is approximately the same as the intravenous dose.

Iron requirements are well known but a means of supplying parenteral iron is still not without hazard. The intravenous preparation has a higher rate of reactions and if it is necessary to use this preparation, emergency resuscitative equipment should be at hand and the patient watched carefully. The intramuscular iron dextran most often is given at a dose of 1 mg iron per day every 7-14 days but individual requirements should be calculated on the basis of need.

It is likely that further research will demonstrate human deficiencies of additional trace elements under certain circumstances and make consideration of requirements for these additional elements mandatory during prolonged parenteral alimentation.

We have analyzed a number of protein solutions which are used in intravenous nutrition (Hambidge, 1971). The results of these measurements are listed in Table 2.

Table 2: Trace elements in protein solutions

	Zinc (ppm)*	Magnesium (mEq/1)	Copper (ppb)+	Iron (ppb)	Chromium (ppb)	Manganese (ppb)
FreAmine 8% (McGraw)	<4	0.012	17.5	41	1.8	3.9
Amino acid 8% (Cutter)	0.125	<4	0.0	—	8.5	0.0
Amino acid 8% mixed for use (Cutter)	0.210	<4	70.	63	1.8	21.
Casein hydrolysate 8% (Cutter)	2.1	1.56	10.	200	4.5	60.
Aminosol 5%, Glucose 5% (Abbott)	0.06	2.76	0.0	32	41.0	104.

* ppm = parts per million   + ppb = parts per billion

It is noteworthy that the Cutter amino acid solution had a much different content of trace elements before and after being mixed with glucose, electrolytes and vitamins for intravenous use. For this reason we measured the elements in most of the solutions that we currently add to the basic protein-glucose mixtures. These data are shown in Table 3.

Table 3: Trace elements in electrolyte and vitamin solutions*

	Manganese	Chromium	Zinc	Iron	Copper	Magnesium
NaCl (3 mEq/ml)	0.06	0	0.160	0	0	0
KCl (2 mEq/ml)	0.10	0	0.025	0	0	0
KH_2PO_4 (1 mEq/ml)	0	0.001	0.290	0	0	high
$MgSO_4$ (4.1 mEq/ml)	0	0.002	0.024	0	0	0
$ZnSO_4$ (1 mEq/ml)	0.017	<0.001	high	0	0	14.5
Ca glucoheptonate (0.9 mEq/ml)	0.110	0.002	0.120	0.575	0	0
$NaHCO_3$ (1 mEq/ml)	0.023	0	0.235	0	0	0
Vitamin K (0.2 mg/cc)	0.007	0.050	0.050	0	0	0
MVI** (USV Pharm)	0.060	0	0	0.010	0	0
B_{12} (μg/ml)	0.009	0.001	0.025	0.002	0	0

*Values are given as μg/100 ml of the usual alimentation solution.
**Multivitamin Infusion product of the USV Pharmaceutical Corp. Composition in 10 ml: Vitamins C, A, D, B_1, B_{12}, E and Panthenol.

The values represent the amounts of each element contained in 100 ml of the final alimentation solution contributed from each source. These data indicate that all the elements measured are significantly lower than the oral or intravenous recommended daily allowances.

Table 4 compares the concentrations of trace elements in the final protein-glucose-electrolyte-vitamin mixture which we are currently using in infants, to the suggested intravenous requirements.

Table 4: Trace elements in alimentation solution* compared to intravenous requirements.

	Manganese	Chromium	Zinc	Copper	Iron
Prepared solution+	0.497	0.112	< 5	0.381	1.57
Calculated intravenous "requirements" per 120 ml solution	10-20	0.2	10-30	15-22	1 mg/day

* Values given as µg per 120 ml.

+ FreAmine mixed to a final concentration of 2.5% with 20% glucose and "standard" amounts of vitamins and electrolytes added, with 1 mg $ZnSO_4$ per 1,000 ml.

Although many of the trace elements are present, they are in very small quantities. Even more important is the fact that the quantity of the elements vary from lot to lot and from one manufacturer to another. Because of this, these values should not be considered as being accurate for all solutions of this type. Even so, these data do suggest that patients given total intravenous alimentation for long periods might be expected to become deficient in certain of the so-called essential trace elements, and because of the variability between different solutions, detailed analysis for trace elements should be performed on a large number of solutions to assure adequate intake and to protect against potentially toxic levels.

We have studied the plasma and urinary changes in one infant who received prolonged intravenous nutrition. The infant weighed 1800 gm and required the removal of all but 20 cm of ileum. Figure 1 illustrates the plasma and urinary zinc concentrations during 30 days of parenteral alimentation. Urinary excretion and serum concentrations of zinc both decreased during the first two weeks of therapy. Beginning on the 25th day, one mg of zinc sulfate was given intravenously each day. Both the serum and urine concentrations subsequently increased significantly. The high serum values would suggest that this dose is in excess of what is required.

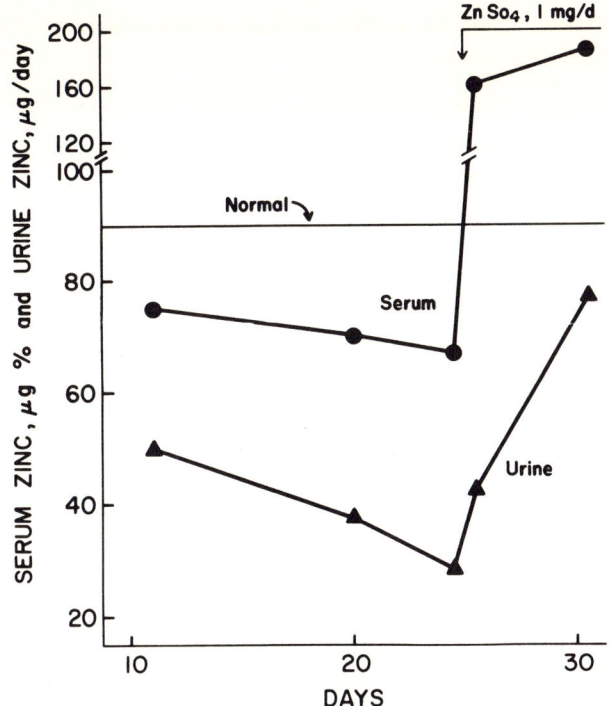

Fig. 1. Effect of "TPA" on serum and urine zinc in an infant.

The copper concentration in this infant showed a decrease over a three week period and returned toward the previous level following a biweekly transfusion of packed red cells. This probably accounts for the fact that the serum copper was never in the range considered to be deficient.

Drs. Filer and Muller, at the University of Iowa, treated an 8 year old girl with total parenteral nutrition for three two-month periods because of a poorly mobile gastrointestinal tract. Each period of treatment was separated by 5 months of oral alimentation. Following the last period of total intravenous nutrition, the same solution was administered via jejunostomy for an additional month. At that time, plasma and hair measurements were made. Table 5 shows that zinc and copper concentrations were significantly low in both plasma and hair. The chromium and manganese concentrations, however, remained in the normal range.

Table 5: Trace elements in parenteral nutrition*

	Zinc	Copper
Plasma	26 µg/100 ml	45 µg/100 ml
	(75 ± 11)	(87 ± 9)
Hair	60 ppm	7 ppm
	(180 ± 40)	(11 ± 2)

* 8 year old girl receiving hydrolysate-glucose solution. See text for details.

Parentheses indicate normal mean values.

Karpel and Peden (1972) have also reported an infant who developed symptomatic copper deficiency despite seven blood transfusions and four plasma transfusions. After 236 days of total parenteral nutrition the serum copper was 9 µg/100 ml. Following 2.5 mg of oral copper per day there was prompt improvement but complete recovery required approximately 3 weeks.

In summary, present knowledge of trace element nutritional requirements indicates that it is mandatory to consider these micronutrients in any subject receiving prolonged total parenteral alimentation. This conclusion is already receiving support from documented trace mineral deficiencies in association with such therapy, and there is no guarantee that such deficiencies can be averted by the administration of blood plasma which contains only extremely small quantities of some of these elements. At the present time, precise information on intravenous requirements for most of these elements is not available. Indeed, it is not known how many of these elements need to be considered. However, some provisional data is available with respect to zinc, copper, chromium, manganese, and of course, iron and iodine. Trace analysis of intravenous solutions at least for these elements should be performed and appropriate additions be made where necessary. Only by this means will the dangers of deficiencies or possible toxicities (Louria et al., 1972) be avoided. Above all, however, further research is urgently required in this field.

VITAMINS

As we are able to satisfy the other nutritional needs of the patient, the ability to assess the intravenous vitamin requirements may become less difficult, since several of the vitamin requirements vary with the amount and ratio of carbohydrate and protein ingested.

Studies concerning the requirements for specific vitamins have been conducted by a number of investigators in our laboratory. They have shown that the biological half-life of certain radioactively labelled vitamins was shorter than expected. Biochemical evidence of deficiency occurred as early as seven days, and abnormalities in electroencephalograms and electrocardiograms appeared in less than three weeks on a B_6 deficient diet (Canham et al., 1966). The half-life of thiamine was between 9 and 18 days in the adult (Ariaey-Nejad, 1970). No normal subjects fed an ascorbate-deficient diet developed petechial hemorrhages as early as the 29th day of depletion, and by the third month all had developed symptoms of ascorbate deficiency (Hodges et al., 1971).

A diet free of riboflavin caused statistically significant changes in red blood cell glutathione reductase activity within seven days (Tillotson, 1972). A diet free of folate caused an elevation in urinary formiminoglutamic acid excretion in four to five weeks (Eichner, 1971). Also, significant changes in jejunal and hepatic glycolytic enzymes occurred in rats before any hematologic changes with a folate deficient diet (Herman et al., 1969). These findings suggest that only a relatively brief period of deficient intake is required to produce biochemical or clinical changes due to a deficiency of some of the water soluble vitamins.

These findings apply to the normal adult during light to moderate activity and not to the rapidly growing infant or older child in various hypermetabolic states whose vitamin requirements may be even greater. For example, folate requirements in the neonate are significantly greater than in the adult (Hoffbrand, 1970; Shojania & Hornady, 1970), and the relatively large amount of folate found in the central nervous system of the newborn mouse (McClain & Bridgers, 1970) may further reflect an increased need for adequate folate intake during this period of rapid brain growth (Shojania & Hornady, 1970). Similar observations have been made with riboflavin and the central nervous system (Rivlin, 1970).

Fat soluble vitamins require a much larger period for depletion in adults. However, low birth weight infants are born with relatively depleted stores of vitamin A in the liver, and supplementation probably should be given if total intravenous alimentation is to be continued for more than two weeks.

Premature infants develop rachitic lesions more readily than term infants. An oral dose of vitamin D as small as 100 International units per day has prevented the development of rickets in term infants and active rickets has been cured with 300 Int. units per day. The recent observations indicating that parathyroid hormone is the trophic hormone necessary for the manufacture of the metabolically active form of vitamin D thereby maintaining calcium homeostasis would suggest that this vitamin should be employed relatively early during total parenteral alimentation in the infant.

The exact function of vitamin E remains unknown but is implicated in the maintenance of normal RBC integrity and hematopoiesis. Human milk contains two to five International units of this vitamin per liter.

The absolute daily allowance of vitamin K has not been established. Adults depleted of vitamin K require only 0.04 mg per kg intravenously to attain normal blood clotting.

Toxicity due to fat soluble vitamins is more of a problem than that due to water soluble vitamins. One 1800 gm infant at our laboratory inadvertantly received 15,000 International units of vitamin A daily for 10 days and developed signs and symptoms of vitamin A toxicity. After the vitamin was discontinued, the abnormalities gradually subsided over an 8 day period.

Dr. Pruitt from the Institute of Surgical Research at Brooke Army Medical Center has observed one adult patient who inadvertently received 30 ml of MVI (US Vitamins, multivitamin preparation for injection) intravenously for 2 weeks, which corresponded to approximately 3,000 International units per day of vitamin D. This patient developed a markedly elevated serum calcium concentration with multiple areas of tissue calcification secondary to vitamin D toxicity (Pruitt, Personal Communication).

At the present time, there is no concrete information as to the intravenous vitamin requirements of the infant. Table 6 shows a comparison of the recommended allowances for vitamins compared with the levels found in one ml of MVI (USV Pharm.). It is apparent that the amount of vitamin A is slightly on the high side while vitamin D is on the low side of the recommended dose. The water soluble vitamins are markedly in excess of the recommended dosage.

In two infants and a 6 year old child who received prolonged intravenous nutrition and MVI (USV Pharm.) at a dose of 1 ml per day, we found that the urinary excretion of unbound thiamine, pyridoxine and riboflavin was extremely high, suggesting that excessive amounts were being used. It is doubtful that excessive amounts of the water soluble vitamins cause any toxic manifestations, however, this is not a certainty in the infant.

Table 6: Vitamin needs for the infant.*

Dietary recommendation per day		MVI** (1 ml)
Niacin	6.6 mg/1000 Cal	10 mg
Thiamine	0.2 mg+	5 mg
Riboflavin	0.4 mg (0.1 mg/kg)‡	1 mg
Pyridoxine	0.2 mg (9 µg/gm protein)	1.5 mg
B_{12}	1.0 to 2.0 µg	—
Ascorbate	35 to 50 mg	50 mg
Folate	0.005 to 0.02 mg	—
Biotin	(?) 1 µg	—
Choline	(?) 7 mg	—
Pantothenic acid	(?) 1 to 2 mg	2.5 mg
Vitamin A	70 to 200 Int. units	1000 IU
Vitamin D	100 to 400 Int. units	100 IU
Vitamin E	.34 to .45 Int. units	0.5 IU
Vitamin K	(?) 0.003 to 0.1 mg	—

*Recommended dietary allowances, Nat. Acad. Sci. (1970) and Fromon, 1967.

**Multivitamin infusion product of the USV Pharmaceutical Corp. (see Table 3).

+ 0.2 to 0.5 mg per 1,000 Cal.

‡ 0.3 mg per 1,000 Cal. or 0.4 mg per 1,000 Cal. for infants over 6 months.

Since large amounts of vitamins were present in the urine of these infants it was recommended that the dose of intravenous vitamins be decreased. The following study, however, suggests that urinary excretion may have little bearing on determining the requirements when the vitamins are given intravenously.

With the idea that we might be able to use the known oral vitamin requirements as a basis to calculate the intravenous requirements, we measured the urinary excretion of B_6, B_2 and B_1 in an eight year old patient (weight 36 kg) receiving total parenteral nutrition (Figure 2).

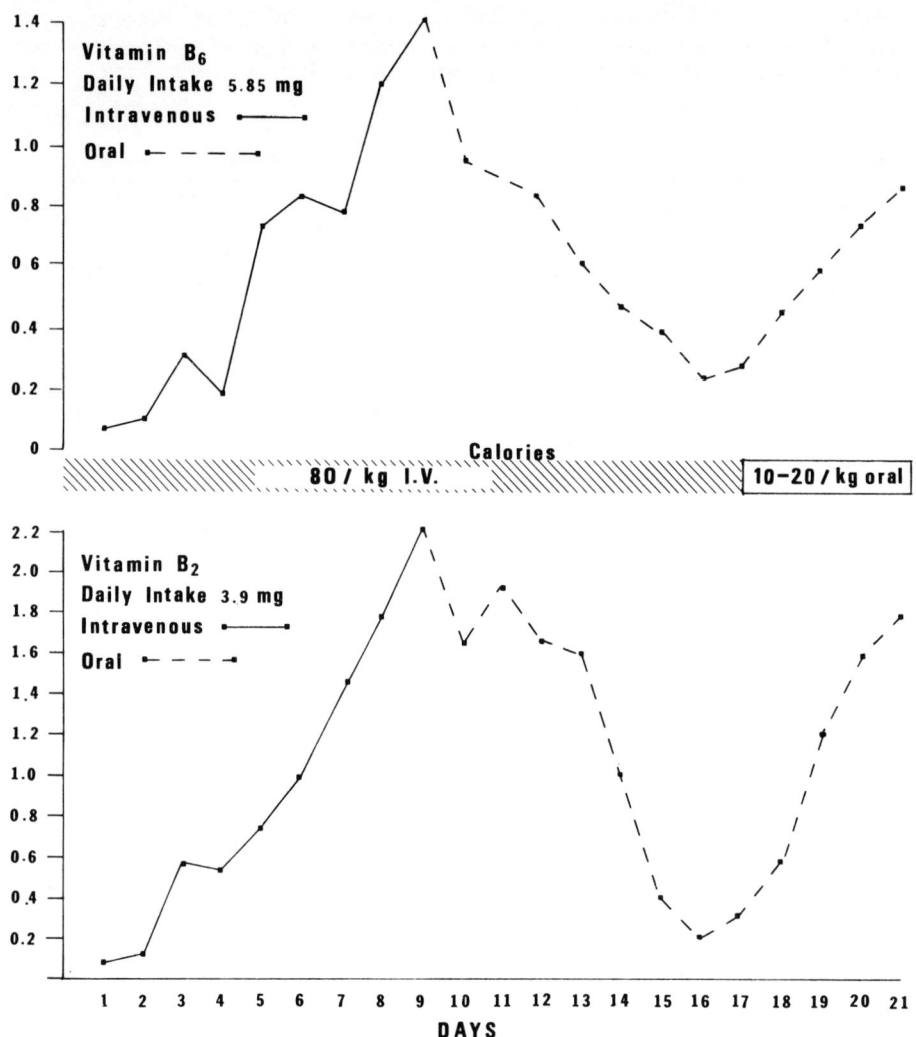

Fig. 2. Urinary excretion of vitamin B_6 and B_2.

During the first 9 days of the study, 5.85 mg of vitamin B_6 was administered intravenously. For the next 8 days, the same dose was given orally. The patient received a constant amount of intravenous calories (80 Cal/kg) and protein for 17 days. There was no intestinal abnormality present and previous studies have shown that approximately 95% of an orally administered dose of B_6 is absorbed. There was a progressive increase in the excretion of B_6 during the period of intravenous vitamins and by the 9th day, almost half of the administered dose appeared in the urine. When the vitamins were given orally, a progressive decrease in the vitamin excretion occurred and after 8 days, the total excretion of both B_6 and B_2 was only slightly higher than at the beginning of the study. Vitamin B_1 excretion was similar to that of B_6 and B_2.

From this initial data it would appear that the vitamins might not have been effectively absorbed from the gastrointestinal tract. However, the same dose of oral vitamins was continued for 4 more days but parenteral nutrition was discontinued and oral feedings were begun. The patient took the vitamins but refused almost all of the diet and the urinary excretion of the vitamins increased almost to the level that was present when the vitamins were given intravenously. These data illustrate the decreased utilization of vitamins in the absence of adequate nutrients and suggest that the vitamins were being absorbed when given orally. With a constant intake of calories and protein the orally administered vitamins appeared to be utilized or catabolized more effectively than when the same dose was given intravenously.

Because of these large variations it appears that the urinary excretion of the vitamins is not an appropriate means of measuring the utilization of at least these 3 vitamins when they are given intravenously. This may in part be due to the fact that a larger portion of the vitamins reach the renal circulation before the hepatic circulation when given intravenously and therefore a larger amount may be filtered and excreted before reaching the liver.

However, the changes in excretion patterns took place gradually, which would not be expected if this were the sole explanation for the high renal losses during intravenous vitamin administration. It is possible that both the gastrointestinal tract and the liver play a more complex role in absorption and metabolism of certain orally administered vitamins than previously recognized. More refined measurements of vitamin adequacy, such as red blood cell transketolase (B_1), glutathiane reductase (B_2), activities or urinary Figlu determinations may give more accurate knowledge of the intravenous vitamin requirements.

In summary: Because of the short time required to develop biochemical evidence of a deficiency of many water soluble vitamins, it would appear that these nutrients should be included from the initiation of any complete parenteral nutrition program. If MVI (USV Pharm.) is used, care should be taken to insure that the patient does not receive excessive amounts of the fat soluble vitamins. Ideally there should be separate preparations of the water and fat soluble vitamins. Unless these solutions are made available for study, the actual requirement for each vitamin will remain difficult to assess. In this regard, it is quite disturbing that recently the FDA, with disregard for the needs of the patients whose lives may be saved by total parenteral nutrition, has placed severe restrictions on the use of intravenous vitamin preparations.

REFERENCES

Ariaey-Nejad, M.R., Balaghi, M., Baker, E.M. and Sauberlich, H.E. Amer. J. Clin. Nutr., 23, 764-778, 1970.

Canham, J.E., Baker, E.M., Raica, N. and Sauberlich, H.E. Proc. 7th Int. Cong. of Nutrition, Vol. 5, Physiology and Biochemistry of Food Components, New York, Pergamon Press, 1966, p. 558.

Cordano, A., Baerti, J.M. and Graham, G.G. Pediatrics, 32, 324-336, 1964.

Davidson, Murray. Pediat. Clin. N. Amer., 17, 913, 1970.

Eichner, E.R., Buergel, N. and Hillman R.S. Amer. J. Clin. Nutr., 24, 1337-1341, 1971.

Fromon, S.J. Infant Nutrition, Philadelphia, W.B. Saunders, 1967, p. 113.

Hambidge, K.M. Anal. Chem., 43, 103-107, 1971.

Hambidge, K.M. Fed. Proc., 1972 (In press).

Hambidge, K.M., Hambidge, C., Jacobs, M. and Baum, D. Ped. Res., 1972 (In press).

Herman, R.H., Stifel, F.B., Herman, Y.F. and Rosensweig, N.S. Fed. Proc., 28, 628, 1969.

Hodges, R.E., Hood, J., Canham, J.E., Sauberlich, H.E. and Baker, E.M. J. Clin. Nutr., 24, 432-443, 1971.

Hoffbrand, A.V. Arch. Dis. Child., 45, 441-446, 1970.

Karpel, J.T. and Peden, V.H. J. Pediatrics, 30, 32-36, 1972.

Levin, R.A., Streeton, D.M.P. and Doisey, R.J. Metab., 17, 114-125, 1968.

Louria, D.B., Joselow, M.M. and Browder, A.A. Ann. Int. Med., 76, 307-319, 1972.

McClain, L.D. and Bridgers, W.F. J. Neurochem., 17, 763-776, 1970.

National Academy of Science, 1969.

Pruitt, B. Personal Communication.

Rivlin, R.S. New Eng. J. Med., 283, 463-472, 1970.

Sandstead, H.H., Burk, R., Booth, G.H. and Darby, W.J. Med. Clin. N. Amer., 54, 1509-1531, 1972.

Shils, M. AMA Symposium on Parenteral Nutrition. Jan 17-19, 1972, pp. 92-114.

Shils, M.E. J. Amer. Med. Assoc., 220, 2110, June, 1972.

Shojania, A.M. and Hornady, G. Pediat. Res., 4, 442, 1970.

Swenerton, H., Shrader, R. and Hurley, L.S. Science, 166, 1014, 1969.

Tillotson, J.A. and Baker, E.M. Amer. J. Clin. Nutr., 25, 425-431, 1972.

Underwood, E.J. Trace Elements in Human and Animal Nutrition, 3rd Ed., New York, Academic Press, 1971.

THE RAPID REHABILITATION OF SEVERELY UNDERNOURISHED CHILDREN

E. M. Widdowson

Dunn Nutritional Laboratory, University of Cambridge

and Medical Research Council

The main concern of this book is with parenteral nutrition, but to help us to find out how much food to give a severely undernourished child through a vein we must consider what an equally undernourished child with a functional gastrointestinal tract will take, and how it will grow if it does so.

In all my experience, whether with children, adults, or young or mature animals, severe undernutrition due to a shortage of food in an otherwise healthy individual is one of the easiest things to treat. The human being or animal is usually ravenously hungry and will eat astonishingly large quantities of food from the word "go". The digestive tract, moreover, which does not seem to lack the necessary digestive enzymes, can deal with the food, and the body weight begins to rise at once. In practice of course, particularly with children, it is not always so simple, for severely undernourished children often have complications of one sort or another, and these have to be treated before the child will react to the food by a rapid gain of weight.

This is illustrated by two African children who were admitted to the Medical Research Council Unit in Kampala under the care of Professor McCance (McCance, 1971). The first, Agnes, weighed only 1.3 kg. when she was born, and 2.5 kg. at 9 months when she was admitted. Her only treatment was to be allowed to eat her fill of a diet providing 8% of its calories as protein. She had no setbacks. She took over 230 kcal. per kg. per day during the 7 weeks she was in hospital and this enabled her to double her weight during this time. Another marasmic child, Kivumbu, was 14 months old when he was admitted and weighed 4.54 kg. - 45% of his expected weight for age. He had a very shaky start with serious attacks of pyrexia and bronchitis during his first 3 weeks, but from then on he did remarkably well on a high calorie diet, taking 170 kcal. per kg. per day and gaining 2 kg. in weight over the whole 4 weeks. He was discharged in the care of his Grandmother who was intelligent enough to give him the food provided for him by the Unit in the quantities prescribed, and his weight continued to go ahead.

Dr. Ann Ashworth, working in Jamaica, has had very similar experiences (Ashworth, 1969a). Her marasmic children, aged 10 - 36 months, with a mean weight of 5 kg. at the beginning, took 160 kcal. per kg. per day over the first 8 weeks, and she believed they might have taken more had it been offered to them. Her diet also provided about 8% of its calories as protein. The Jamaican children gained 3.7 kg. during the first 8 weeks they were in hospital. Dr. Ashworth reckons that the mean rate of gain in weight of her children was 15 times as fast as the normal growth rate of children of the same age and 5 times as fast as that of younger children of the same weight.

How much the gain in weight depends on the calorie intake is illustrated in Fig. 1 (Rutishauser & McCance, 1968). This shows

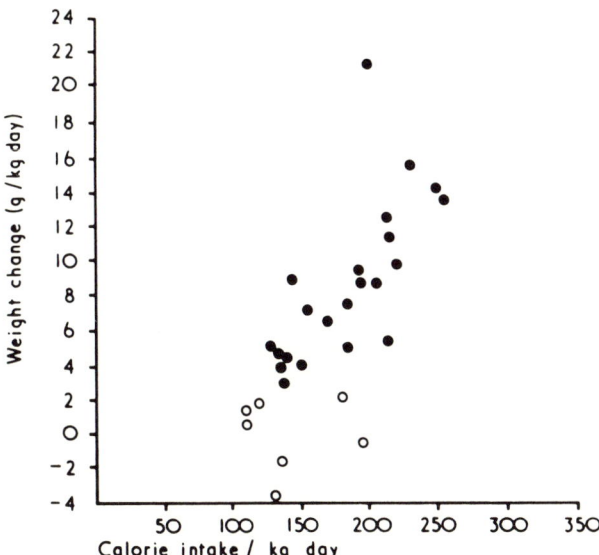

Fig. 1 Change in weight and calorie intake of undernourished children given diets providing 4 g. protein per kg. day.
● Weight change associated with diarrhoea or infection.
O Weight change uncomplicated by infection.
(From Rutishauser & McCance, 1968)

the change in weight plotted against the calorie intake of a number of undernourished Ugandan children given a diet providing 8 - 9% of its calories as protein, or 4 g. protein per kg. per day. No advantage is to be gained by giving more protein than this. The solid circles represent children who had no infection and the rise in the gain of weight with increase in the intake of calories is clearly shown. The open circles represent children who had diarrhoea or other infection and they gained less weight on a similar number of calories. Calorie intakes of 230, 170 or even 160 per kg. per day are far above those recommended for or taken by normal children. Table 1 shows the recommended allowances for children up to 1 year published in Britain (Department of Health and Social Security, 1969); the American values are very similar, falling from 120 kcal. per kg. per day at birth to 100 kcal. per kg. per day at 1 year (Food and Nutrition Board, 1968). It was once thought that 100 kcal. per kg. per day were sufficient for undernourished children and they were given no more, but recent studies on children have made it clear how much more the severely undernourished child needs and will willingly take.

Table 1: Recommended Daily Intakes of Energy for Infants

(from: Recommended Intakes of Nutrients for the United Kingdom, 1969, Department of Health and Social Security)

Age range	Body weight kg	kcal/kg
Birth up to 3 months	4.6	120
3 up to 6 months	6.6	115
6 up to 9 months	8.3	110
9 up to 12 months	9.5	105

Those who have worked in the Medical Research Council Unit in Uganda believe that unless the diet given to a child who has been severely undernourished for a long time provides more than 150 kcal. per kg. per day it will not enable the child to return to its proper growth curve (Rutishauser & McCance, 1968). The child will grow on fewer calories but not catch up, and it will necessarily, therefore, remain underweight for a long time and possibly for the rest of its life. The younger the child, and the more underweight it is, the more calories per kg. it will need.

The diets for marasmic children both in Uganda and Jamaica provide 8 - 9% of their calories as protein, over 60% as fat and 25 - 30% as carbohydrate. The one used in Uganda is set out in

Table 2, which gives all quantities per kg. body weight per day (Staff, 1967). Children having this diet need additional iron and trace elements, and they may need vitamin supplements.

Table 2: Diet Used Successfully for Rapid Refeeding of Severely Undernourished Children in Uganda

Amounts per kilogram body weight per day

Full cream milk powder	16 g.
Sucrose	7 g.
Cotton seed oil	10 g.
Water	167 g.
This provides	
Calories	200
Protein	4.0 g.
Fat	14.3 g.
Carbohydrate	13.2 g.
Potassium	5.2 meq
Sodium	2.8 meq
Magnesium	1.5 meq

Why do undernourished chilren need so many calories and where do the calories go? Ashworth (1969b) has provided one important reason for this. She measured the oxygen consumption of children recovering from marasmus and found that it rose more and stayed high longer after each meal in these children than it did in the normal child. She described this as the energy required for their rapid growth, and more particularly, the energy cost of protein synthesis. These children required more energy for the laying down of new tissue than normal children because they were growing so rapidly and laying down such large amounts of tissue. Ashworth postulated that they grow in spurts after each meal, and not at a regular rate throughout the 24 hours.

Undernourished children who have to be fed by vein require no less food than those who have a normal digestive tract, and, moreover, the stress to which they have been subjected may put up their energy requirements even higher. Parenteral feeding which provides 125 kcal. per kg. per day may well be sufficient for the child who is not far below normal weight when the treatment begins, but may not be adequate for growth of severely undernourished children. Measurements of oxygen consumption of parenterally fed children would also be useful. A comparison of children near normal weight with others severely undernourished, all being fed continuously round the clock, would be particularly rewarding.

REFERENCES

Ashworth, A. Growth rates in children recovering from protein-calorie malnutrition. Brit. J. Nutr., 1969a, 23, 835 - 845.

Ashworth, A. Metabolic rates during recovery from protein-calorie malnutrition: the need for a new concept of specific dynamic action. Nature, Lond., 1969b, 223, 407.

Department of Health and Social Security. Recommended intakes of nutrients for the United Kingdom. Reports on Public Health and Medical Subjects, No. 120. HMSO : London, 1969.

Food and Nutrition Board. Recommended dietary allowances, 7th ed. National Academy of Sciences : Washington, D. C., 1968.

McCance, R.A. Malnutrition in Uganda. Indian J. med. Res., 1971, 59, Suppl. 132 - 142.

Rutishauser, I.H.E. and McCance, R.A. Calorie requirements for growth after severe undernutrition. Arch. Dis. Childh., 1968, 43, 252 - 256.

Staff, T.H.E. Treatment of severe kwashiorkor and marasmus in hospital. East African Med. J., 1967, 45, 399 - 406.

TECHNIQUE OF TOTAL PARENTERAL NUTRITION IN INFANTS

Stanley J. Dudrick, Bruce V. MacFadyen, Robert W. Winters

The Program in Surgery, The University of Texas Medical

School at Houston, Texas Medical Center, Houston, Texas

Normal growth and development in any animal species nourished entirely by vein was achieved for the first time in Beagle puppies in the Harrison Department of Surgical Research of the University of Pennsylvania in 1965 (Dudrick et al., 1967; Dudrick et al., 1970). Repeated demonstrations of normal growth and development in subsequent litters of puppies as well as positive nitrogen balance in adult surgical patients suggested the feasibility and safety of providing all nutrients entirely by vein to newborn human infants. Based on knowledge of the daily average oral pediatric nutritional requirements, on information obtained from the puppy studies, and on serial serum and urine measurements during infusions of dextrose, nitrogen, vitamins, minerals and electrolytes in man, a solution providing theoretically adequate intravenous requirements of the major nutrients for newborn infants was developed (Wilmore et al., 1968; Dudrick et al., 1969; Wilmore et al.). Initially, 4 gm of protein in the form of fibrin hydrolysate were given per kilogram body weight daily. Subsequently, normal growth and development have been achieved in infants receiving 2½ to 3 gm of intravenous protein hydrolysates/kg/day. Historically, the initially higher dose of nitrogen substrates was given to insure delivery of maximum quantities of protein moieties for tissue synthesis, and such dosages may still be required to promote normal growth and development in pediatric patients having pathologic conditions associated with marked catabolism.

Because of the current unavailability of intravenous fat emulsions for routine clinical administration in this country, this important constituent of a truly balanced dietary ration is necessarily omitted. Therefore, all of the non-nitrogenous calories

Table 1: Comparison of Average Daily Pediatric Requirements Per Kilogram Body Weight

	Oral Recommended		Intravenous Hyperalimentation		
Protein	2.5	gm	2.5- 4	gm	
Calorie	115	Kcal	115 -135	Kcal	
Water	150	ml	125 -145	ml	
Sodium	46	mg	100	mg	(4-5 mEq)
Potassium	58	mg	156 -195	mg	(4-5 mEq)
Chloride	150	mg	150	mg	(4 mEq)
Calcium	218	mg	72	mg	(3-4 mEq)
Phosphorus	218	mg	58	mg	(5-6 mEq)
Magnesium	60	mg	25	mg	(2 mEq)
Iron	6	mg	0.02	mg	
Copper	0.07	mg	0.022	mg	
Cobalt	-		0.014	mg	
Manganese	0.2	mg	0.04	mg	
Zinc	0.3	mg	0.04	mg	
Iodine	0.07	mg	0.015	mg	

which are required to meet energy needs and to promote protein synthesis must be provided in the form of carbohydrates, polyols or ethyl alcohol. Because polyols, such as sorbitol and xylitol, are not approved by the Food and Drug Administration for parenteral use in human beings, and because ethyl alcohol may cause subclinical impairments of cellular potential and function, especially in the central nervous system, they are not recommended for use in parenteral hyperalimentation regimens. Thus, the only available, practical and safe source of intravenous energy in the United States is carbohydrate. As no significant advantage is to be gained, and some adverse metabolic effects have been attributed to the use of fructose or invert sugar, dextrose is the carbohydrate fuel of choice in pediatric total parenteral nutrition at this time. When dextrose, which yields 3.7 calories per gram, or dextrose monohydrate, which yields 3.4 calories per gram, is given in sufficient quantity to meet energy requirements, the solution must be hypertonic if the limits of water metabolism are not to be exceeded. Moreover, the dextrose must be given continuously throughout each day if maximum assimilation and minimum renal excretion of the sugar is to be achieved. The usual range of dextrose utilization in newborn infants is 0.4 to 1.2 gm/kg/hour, and in full term neonates, the maximum dose level can be reached within the first few days. Relative glucose intolerance can occur in the immediate postpartum period, especially in premature infants. However, rarely in our experience has exogenous insulin administration been required to induce growth or development in infants receiving intravenous hyperalimentation, and its routine use is condemned.

Table 2 compares the intravenous dosages of vitamins administered in the pediatric hyperalimentation solution with the oral recommendations. Vitamins A through pantothenic acid have been given in the fixed ratios which are present in the only commercially available parenteral vitamin mixture which contains both fat and water soluble vitamins. The limiting vitamin in this mixture for infants is vitamin D. This vitamin is given in a dose of 300 to 400 international units per day in order to prevent the development of rickets. Of necessity, the remainder of the vitamins are thereby given in somewhat excessive quantities because of the fixed ratios in the commercial product. To date, however, hypervitaminosis has not been reported. Vitamin K, folic acid, and vitamin B_{12} are given individually in approximately the daily recommended doses. During the initial stages of pediatric parenteral hyperalimentation, it was assumed that the intravenous dose of vitamin D would be somewhat less than the oral recommended dose because a major function of vitamin D is to aid in the absorption of calcium and phosphorus across the intestinal mucosa. A reduction in the intravenous dose of vitamin D to 100 to 200 international units for a few infants, however, produced decalcification, particularly of the tibial plateau, and the development of a rachitic rosary at the costochondral junctions within three weeks. Reestablishment of parenteral vitamin D administration at a dosage level of 400 international units per day was followed promptly by regression of the rachitic clinical manifestations.

A unit of pediatric hyperalimentation solution capable of providing the daily nutrient requirements for the average newborn infant can be formulated readily in the United States from commercially available products. The combination of 400 ml of 5% protein hydrolysate or crystalline amino acids with 250 ml of 50% dextrose produces a solution which contains approximately one calorie per ml.

Table 2: Comparison of Average Daily Pediatric Vitamin Requirements

	Oral Recommended		Intravenous	
Vitamin A	1,500	IU	3,000-4,000	IU
Vitamin C	30	mg	150-200	mg
Vitamin D	400	IU	300-400	IU
Vitamin E	---		1.5-2.0	IU
Thiamine	0.4	mg	15-20	mg
Riboflavin	0.6	mg	3-4	mg
Pyridoxine	0.25	mg	4.5-6.0	mg
Niacin	6	mg	30-40	mg
Pantothenic Acid	---		7.5-10.0	mg
Vitamin K	1.5	mg	1.0-1.5	mg
Folic Acid	0.35	mg	0.5	mg
Vitamin B_{12}	1	mcg	1	mcg

In order to complete the nutrient solution, however, approximately 75 ml of additives are made to the base solution prior to administration to the patient. All nutrient solutions should be prepared under strict aseptic conditions by a specially trained pharmacist, nurse or technician in a laminar-flow, filtered air hood in order to minimize the risks of contamination with micro-organisms.

The basic guidelines for safe intravenous hyperalimentation in infants include accurate daily measurements of body weight, hourly water balance, fractional urine sugar concentration every six hours, serum glucose and electrolytes daily until stable and every two or three days thereafter, and complete blood count, blood urea nitrogen and blood ammonia weekly. Head circumference and body weight should also be measured weekly. It is advisable to evaluate hepatic function, serum calcium, phosphorus, magnesium and proteins initially and every one to three weeks thereafter during high calorie intravenous therapy. Occasional determinations of serum osmolality and vitamin levels and urine specific gravity, osmolality, acetone, electrolytes and amino acids may be helpful in monitoring the

Table 3: Unit Preparation of Pediatric Nutrient Solution

Base Solution

400 ml { 5% Dextrose / 5% Protein Hydrolysate }   160 Cal { 20 gm Dextrose / 20 gm Protein Hydrolysate }

+

250 ml 50% Dextrose   500 Cal

650 ml   660 Cal

+

~75 ml   Additives

~725 ml Final Solution

Infusion rate: 145 ml/kg/day = 130 Cal/kg/day

---

metabolic status of some infants. Periodic measurements of arterial and central venous pressures, blood gases and pH may also be indicated in the management of critically ill patients with significant cardiovascular, respiratory or metabolic derangements.

Because of the marked hypertonicity of the concentrated nutrient solution, which exerts an osmotic pressure of approximately 1800 to 2400 milliosmoles depending upon the additives, it must be delivered at a constant rate throughout the entire 24-hour day into a large diameter, high flow vessel such as the superior vena cava. A fine silicone rubber or polyvinyl catheter having an external diameter approximately that of an 18 gauge needle is inserted by cut-down into a common facial, external jugular or internal jugular vein using strict aseptic and antiseptic surgical technique (Dudrick et al., 1968). The catheter is directed into the mid-superior vena cava and is fixed in place with ligatures. The length of catheter which must be inserted is equal to the distance from the cut-down site in the neck to the second intercostal space. The catheter must not be advanced into the heart or into the inferior vena cava because of the increased incidence of clotting and endocarditis which has been reported in patients in whom this has been done (Asch et al., 1972).

Table 4: Unit Preparation of Pediatric Nutrient Solution

		Additives		
Sodium	20 mEq	Sodium Chloride (2 mEq/ml)	10	ml
Potassium	25 mEq	Potassium Acid Phosphate (2 mEq/ml)	13	ml
Phosphorus	25 mEq			
Calcium	20 mEq	Calcium Gluconate 10% (0.45 mEq/ml)	44	ml
Magnesium	10 mEq	Magnesium Sulfate 50% (4 mEq/ml)	2.5	ml
Multiple Vitamin Infusion			4	ml
Vitamin K, Vitamin B_{12}, Folic Acid, Iron		Added to solution daily or weekly or given intramuscularly intermittenly	1	ml
Trace Elements		Added to solution daily or given as 10 ml plasma/kg twice weekly	1	ml
			~ 75	ml

After the catheter has been secured in the neck, it is directed subcutaneously to emerge through the parietal scalp behind the ear either by threading it onto a modified Kirschner wire or by directing it through the lumen of a large-bore Vim Silverman needle. As the needle is removed gently, the catheter is left in place in a long subcutaneous tunnel with minimum operative trauma to the infant. Thus, the external portion of the catheter is remote from the searching hands of the infant, and the risk of sepsis is reduced by separating the skin exit site from the venous entry site of the catheter. The catheter is sutured to the scalp, and the neck wound is closed with fine sutures. Antimicrobial ointment is placed at the catheter exit site, and a sterile occlusive dressing is applied. A 0.22 micron membrane filter having a diameter of 2.5 cm is inserted in-line between the catheter and the intravenous administration tubing. Before administering the hypertonic solution, catheter tip location within the superior vena cava must be determined by chest fluoroscopy or roentgenography of the infant.

TECHNIQUE OF TOTAL PARENTERAL NUTRITION IN INFANTS

If the catheter is not radio-opaque, 0.5 ml of water soluble contrast material should be injected into the catheter. After correct catheter placement has been confirmed, the hypertonic solution is continuously and finitely delivered to the infant by means of a constant infusion pump. If metabolic balance studies are to be done, the infant should be maintained on a metabolic bed to facilitate accurate collection of all external secretions.

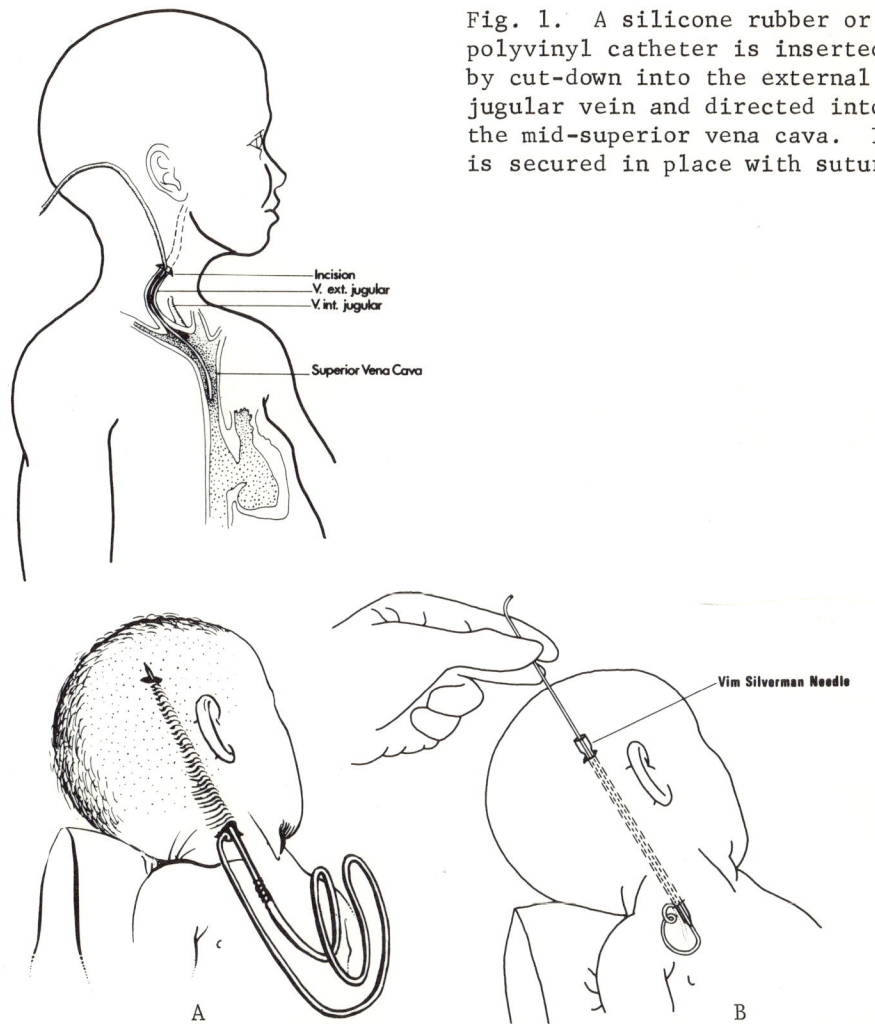

Fig. 1. A silicone rubber or polyvinyl catheter is inserted by cut-down into the external jugular vein and directed into the mid-superior vena cava. It is secured in place with sutures.

Fig. 2. The catheter is directed subcutaneously using a Kirschner wire (A) or alternatively through a Vim Silverman needle (B) to emerge through a small incision in the parietal scalp behind the ear. With minimum trauma to the infant, the catheter is thereby placed in a long subcutaneous tunnel.

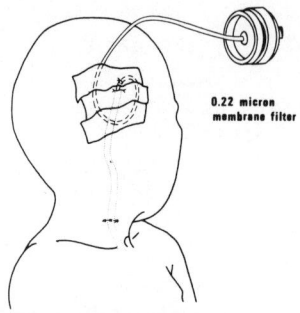

Fig. 3. The catheter is sutured to the scalp at its exit site, antimicrobial outment is applied, and a sterile gauze dressing is fixed to the skin with adhesive tape.

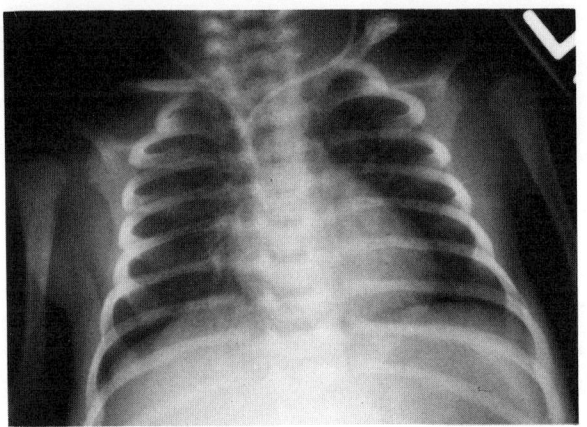

Fig. 4. This roentgenogram shows two central venous catheters in the superior vena cava. The right-sided one is in correct position. The left-sided one is not, and it was removed after proper placement of the right-sided catheter was confirmed in this film.

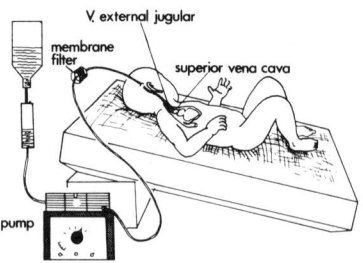

Fig. 5. Hyperalimentation solution is delivered to the infant continuously and accurately by means of a constant infusion pump. A 0.22 micron membrane filter is inserted between the central venous feeding catheter and the intravenous administration tubing. The infant is maintained on a metabolic bed in order to provide accurate collection of all urine and other wastes.

In order to insure maximum safety to the patient, it is strongly recommended that any institution in which intravenous hyperalimentation is practiced should have a designated and qualified hyperalimentation or total parenteral nutrition team. The minimum team should consist of an attending physician who is well versed in nutrition and metabolism, a conscientious and interested house officer or research fellow, a pharmacist trained in aseptic solution formulation, and a technician or nurse. Other members of a complete hyperalimentation team include a nutritionist, clinical pathologist or biochemist, physical therapist, sociologist, and psychiatrist. It had been demonstrated all too often in the literature (Ashcraft et al., 1970; Bernard et al., 1971; Curry et al., 1971) that failure to comply with the principles and procedures outlined by our team for the execution of safe and effective intravenous hyperalimentation will result in an inordinately high rate of metabolic, mechanical and infectious complications. In order to minimize the development of inflammation and infection at the catheter-skin exit site, the dressing is changed at least three times a week; the skin around the catheter is treated with tincture of iodine or other antiseptic solutions; and fresh antimicrobial ointment is applied. A sterile dressing is fixed to the skin with tincture of benzoin and adhesive tape, and the intravenous delivery tubing is changed. All connections in the intravenous line are secured with tape to prevent accidental disengagement with resultant contamination, fluid loss, or air embolism. The importance of conscientious, meticulous and regular catheter care cannot be overemphasized.

The indwelling central venous catheter should be maintained as an intravenous life-line. That the catheter should be used exclusively for delivery of the intravenous nutrient solution is inherent in this philosophy. The temptation to withdraw blood via the catheter, to use the catheter for frequent central venous pressure monitoring, to inject bolus medication via the catheter, or to use the catheter for blood or blood constituent administration must be repressed. Three-way stopcocks within the delivery system must be condemned, because maintenance of the sterility of any intravenous system containing a three-way stopcock is virtually impossible.

Should the infant exhibit an elevated temperature or other signs of infection or sepsis, the physician should promptly and thoroughly evaluate the fever of unknown origin. Accordingly, the ears, upper respiratory tract, lungs, urinary tract, and any surgical wounds must be completely examined. If no source of systemic infection can be detected, the intravenous tubing and bottle or bag of nutrient solution should be cultured and replaced. If the patient does not improve in one or two hours, the catheter should be incriminated empirically as the source of sepsis, and it should be removed and cultured immediately. Usually 24 to 48 hours are allowed to elapse prior to insertion of another central venous feeding catheter on the contralateral side. In the interim, a scalp vein needle should be placed into a peripheral vein, and a solution containing isotonic or slightly hypertonic dextrose should be infused in order to prevent "rebound" hypoglycemia. In some patients, it may be necessary to insert another central venous catheter immediately because of the desperate nature of the patient's condition and nutritional status.

More than 150 infants have been given long-term total intravenous nutritional support by our group. Eighteen of these patients have had either a ruptured omphalocele or gastroschisis. With the use of intravenous hyperalimentation as an adjunct to their surgical management, the mortality rate in these 18 consecutive infants has been zero. Previous experience in treating patients with these congenital defects in our institutions prior to the clinical application of intravenous hyperalimentation yielded a 60 to 80 percent mortality rate. Hence, the value of adequate parenteral nutritional support and bowel rest in the management of these major congenital conditions is obvious.

Clinical Example #1

The initial physical examination of a 2780 gm female infant, following a normal pregnancy and spontaneous delivery, revealed the presence of a ruptured omphalocele. The extruded small bowel and right colon were covered with a fibrinopurulent exudate and were markedly edematous and dilated. No other congenital abnormalities were apparent. The child was taken to the operating room where the irreducible viscera were covered with an envelope fashioned from silastic-impregnated dacron cloth which was sutured to the fascial edges of the abdominal wall defect.

Fifteen hours after the operation, a polyvinyl catheter was inserted into the superior vena cava via the right external jugular vein. Intravenous hyperalimentation was instituted in order to "rest" the inflamed, edematous, dilated bowel and simultaneously to provide nutrients for growth of the infant with the intention that the bowel would soon regain its "right of domain" within the abdomen. By the fifteenth postoperative day, the peritoneal cavity was able to accommodate all of the bowel, and on the sixteenth day, the infant passed feces per rectum. Accordingly, oral feedings were instituted as a supplement to the hyperalimentation solution and were gradually increased to tolerance as the intravenous nutrient regimen was concomitantly decreased. On the twenty-eighth postoperative day, she underwent a second operation in which the dacron sheet was excised from the abdominal wall, and the central venous feeding catheter was removed to minimize the risk of sepsis. For three days she was fed by peripheral veins with standard pediatric intravenous solutions. On day 30, she experienced another spontaneous bowel movement, and the following day, oral feedings became her sole source of nourishment. By the fortieth postoperative day, the base of the wound was completely filled with granulation tissue, and epithelialization was progressing spontaneously. By the forty-ninth postoperative day, the wound was completely healed by secondary intention. The child is now five years of age and has no apparent residual health problems.

After an initial weight loss which is common to all newborn infants as they mobilize excess body water postpartum, this infant gained weight normally at a rate of approximately 30 gm/day. The ability to metabolize an intravenously-administered dextrose load differs somewhat in each infant. It is recommended that the initial parenteral nutritional efforts after birth be carried out with standard pediatric 5% dextrose and 0.25% saline solutions. Following insertion of the superior vena cava catheter, the concentration of dextrose should be increased gradually over the ensuing few days until full strength hyperalimentation solution can be tolerated. Acceptable assimilation of the dextrose load is determined by serial measurements of serum and urine glucose levels. If hyperglycemia and glycosuria occur, an osmotic diuresis will ensue, and the infant may become rapidly and severely dehydrated. The adaptation of the infant's metabolic pathways to tolerate the ever increasing dextrose load has not been convincingly identified. In one infant in whom serum insulin levels have been recorded, a marked rise in the plasma insulin level was initially noted when 20% dextrose solution was infused. The plasma insulin level, however, regressed to normal after five days of infusion, and the infant remained normoglycemic (Das et al., 1971).

After intravenous hyperalimentation was discontinued and oral feedings were begun in the infant herein described, she spontaneously ingested a diet comparable in calories and nitrogen to that which had been provided by vein.

Clinical Example #2

Following an uneventful pregnancy and normal spontaneous delivery, a female infant weighing 2300 gm was born on July 16, 1967. Severe protracted vomiting developed on the second day after birth. Roentgenographic examination with barium contrast enema revealed an obstruction at the distal sigmoid colon. A dilated fluid-filled stomach and duodenum with an absent gas pattern distal to the ligament of Treitz were also noted. Initially nutrition was provided with standard periatric 5% dextrose and 0.25% saline solutions infused via peripheral veins. Exploratory laparotomy revealed massive small bowel atresia from the ligament of Treitz extending to the terminal 3 cm of ileum. There was an atretic segment 2 cm in length in the mid-transverse colon as well as a high rectal stricture.

The atretic small bowel was resected and the bulbous end of the duodenum was anastomosed end-to-side to the terminal 3 cm of ileum. The splenic flexure of the colon was anastomosed to the hepatic flexure, and a sigmoid loop colostomy was performed to bypass the rectal obstruction. A gastrostomy tube was inserted for decompression. The infant was then fed by peripheral veins with plasma, blood, and 10% dextrose and amino acid solutions containing vitamins and minerals. On this regimen, however, her weight dropped from 2300 gm at birth to 1816 gm in 18 days. She became extremely hypometabolic, manifested by pulse rates of 60 to 80 per minute, respiratory rates of 12 to 14 per minute and a core body temperature of 96°F even while in a 101°F isolette. A polyvinyl feeding catheter was inserted into her superior vena cava via an external jugular vein, and continuous infusion of pediatric hyperalimentation solution was begun.

Within 45 days after the institution of intravenous hyperalimentation, the patient had gained weight from 1816 gm to 3405 gm. She also increased in body length by 5.5 cm, head circumference by 6.5 cm and chest circumference by 8.5 cm. The infant was maintained in positive nitrogen, sodium, and potassium balances throughout this period of time. The importance of balance studies is indicated by the calcium and phosphorus metabolism in this infant. Insufficient calcium and phosphorus were provided to produce positive balances until day 20. Hypophosphatemia and hypocalcemia will regularly result if phosphorus and calcium are not provided in adequate amounts in the nutritional solution daily (Ruberg et al., 1971). It is noteworthy that positive phosphorus balance preceded positive calcium balance in this infant in a fashion similar to that which usually occurs in infants fed by mouth.

This infant was maintained on intravenous hyperalimentation for 21 months. An upper gastrointestinal series performed at one year of age revealed marked dilatation of the duodenum with increased mucosal folds and prolonged transit time. The length of bowel had not increased any more than one would expect from natural growth, but the circumferential increase was marked.

In an attempt to determine the prognosis of this infant and subsequent infants with short gut syndrome, an experimental model was designed utilizing Beagle puppies. Three groups of puppies were studied. Ninety percent of the small bowel was excised in two of the groups, with the third group serving as non-operated litter mate controls. Arbitrarily, one set of the resected animals was fed for one month with parenteral hyperalimentation and given nothing by mouth. The other set of resected animals was allowed to eat ad lib by mouth following three days of postoperative intravenous nutritional support. After the 30-day experimental period, the animals who were nourished via intravenous hyperalimentation were allowed to eat horse meat, cereal and water ad lib in the same manner as their resected and non-resected controls for the first year of life. Near-normal growth and development were achieved in all of the bowel-resected puppies, who were fed with intravenous hyperalimentation for the first 30 postoperative days. In contrast, the initially orally-fed, bowel-resected litter mates failed to thrive, had an increased mortality rate, and achieved only half-normal size at the end of one year. In the animals fed for 30 days with parenteral hyperalimentation, the duodenal villus height, mucosal thickness, and total bowel wall thickness were approximately twice that of the non-resected control animals and significantly greater than that of the animals which were bowel-resected, but fed initially by mouth. These data strongly support the Flint (Flint, 1912) hypothesis that bowel adaptation is secondary to increased mucosal absorptive surface area rather than to increased transport across the individual cells.

An alternative method of total parenteral nutrition in infants utilizes a soybean oil emulsion as a major source of calories. Infusing a nutrient mixture which closely resembles breast milk in its basic composition, Borresen and his associates (Borresen et al., 1970) have maintained 32 postoperative newborn infants for periods of 3 to 40 days. By providing 4 gm/kg/day of the fat emulsion, the sugar concentration and hypertonicity of the solution can be decreased without reduction of total calories. Thus the fat and the crystalloid components of the parenteral nutrient regimen can be infused continuously over 24 hours via a Y-tube into peripheral veins, and the risk of long-term central venous catheterization can be avoided. Further investigations are indicated in this important area if parenteral nutrient regimens are to be truly complete, but will be difficult to accomplish in this country until the present restrictions on the clinical use of intravenous lipid emulsions are relaxed.

In summary, the techique of intravenous hyperalimentation is the first parenteral feeding technique that has allowed normal growth and development for prolonged periods of time in animals and man, and remains the only means by which to provide such support in countries like the United States of America, where intravenous fat emulsions are not generally available for clinical use. It is not only life-saving in the management of many pediatric congenital catastrophies, bowel dysfunctional syndromes, and, as recently reported, extreme prematurity, but also offers the scientist a unique opportunity to study in the laboratory and in man many aspects of nutrition, metabolism, pathophysiology, and applied biochemistry in a manner heretofore impossible to achieve.

References

Asch, M.J., Huxtable, R.F. and Hays, D.M. Arch. Surg., 104, 434, 1972.

Ashcraft, K.W. and Leape, L. J.A.M.A., 212, 454, 1970.

Bernard, R.W., Stahl, W.H. and Chase, R.M. Ann. Surg., 173, 191, 1971.

Borresen, H.C., Coran, A.G. and Knutrud, O. Ann. Surg., 172, 291, 1970.

Caldwell, M.D., Jonsson, H.T. and Otherson, H.B. J. Ped., 81, 894, 1972.

Curry, C.R. and Quie, P.G. N. Engl. J. Med., 285, 1221, 1971.

Das, J.B., Filler, R.M., Rubin, V.G. and Eraklis, A.J. J. Ped. Surg., 5, 127, 1970.

Dudrick, S.J., Rhoads, J.E. and Vars, H.M. in Fortschritte der Parenteralen Ernahrung. Symposium of the International Society of Parenteral Nutrition in 1966. Pallas Verlag, Lochham bei Munchen, West Germany, 1967.

Dudrick, S.J., Steiger, E., Wilmore, D.W. and Vars, H.M. Lab. Anim. Care, 20, 521, 1970.

Dudrick, S.J. and Wilmore, D.W. Hosp. Prac., 3, 65, 1968.

Dudrick, S.J., Wilmore, D.W., Vars, H.M. and Rhoads, J.E. Ann. Surg., 169, 974, 1969.

Flint, J.M. Bull. Hopkins Hosp., 23, 127, 1912.

Ruberg, R.L., Allen, T.R., Goodman, M.J., Long, J.M. and Dudrick, S.J. Surg Forum, 22, 86, 1971.

Wilmore, D.W. and Dudrick, S.J. J.A.M.A., 203, 860, 1968.

Wilmore, D.W., Dudrick, S.J., Daly, J.M. and Vars, H.M. Surg., Gynecol., Obstet., 132, 673, 1971.

Wilmore, D.W., Groff, D.B., Bishop, H.C. and Dudrick, S.J. J. Ped. Surg., 4, 181, 1969.

POSTOPERATIVE PARENTERAL FEEDING OF NEONATES: PERIPHERAL VEIN

INFUSION TECHNIQUE, FAT ADMINISTRATION AND METABOLIC STUDIES

H.C. Børresen, R. Bjordal, and O. Knutrud

Departments of Clinical Chemistry and Paediatric Surgery

Rikshospitalet, Oslo 1, Norway

Introduction

Indications for total intravenous feeding. Paediatric Surgical Departments take care of an increasing number of newborn children with life threatening malformations and other complications precluding oral feeding for prolonged periods. The longer this period is, the more the survival of the patient depends on the feasibility of balanced total parenteral feeding. Furthermore, parenteral feeding, insofar as it sustains normal growth, counteracts the permanent reduction in brain size which may result from protein-calorie-malnutrition in early life (1).

Our practice is to institute fully adequate total parenteral feeding programs in those cases where sufficient oral feeding is not likely to start until between the 4th and 6th postoperative day. Thus patients with intestinal resections, anastomoses, gastrochisis, etc., are put on the complete parenteral feeding regime as soon as their clinical condition allows (circulation, respiration, renal function). Patients with diaphragmatic hernia, oesophagus atresia, etc., are borderline cases which are usually treated with simplified fluid-electrolyte infusion programs.

Patient material and clinical experience. Between 1959 and 1972, 844 newborns have been operated upon in the Paediatric Surgical Department at the State University Clinic in Oslo. In 1970 and 1971, the figures were 129 and 125 respectively. The average survival rate has been approximately 80%. In 1970, 37 newborns with especially complicated surgical problems were treated with total parenteral nutrition for periods from 3 to 35 days. Eight of these died. Thus 60 newborns and infants were fed intravenously for 9 days on the average, with a mortality of 16%. We regard this figure to be promising in such a high risk group. Our group of 9 patients with gastrochisis illustrates our point particularly well: 6 of these 9 patients survived. They were given parenteral feeding from 8 to 34 days. It is well known that such patients nearly always starved to death in the past.

Purpose and content of our research program. Through the last 6 years we have been constantly using and improving routine parenteral feeding programs for newborns and infants. Our aim is to establish optimal postoperative requirements of newborns on total parenteral feeding. The effects of the feeding programs have been investigated by multiple simultaneous balance studies. These data are evaluated against the previous studies of one of us (O.K.) on the metabolic response of newborns to operation and inanition (2), and compared to reports in the literature on balance studies during oral feeding (3). We have recently extended our methods to include measurement of serum triglycerides and amino acids.

Metabolic peculiarities of the newborn

The judicious design and application of parenteral feeding programs in the care of newborns must be based on considerations of physiological and biochemical peculiarities of neonates. The neonate reacts to the combined stress of surgery and inanition in much the same way as the adult patient (2). It is not universally appreciated, however, that equally excessive catabolism and metabolic exhaustion can result from prolonged and difficult labor (2). Intrauterine malnutrition (small-for-dates) can also be expected to decrease the infant's tolerance to stress and further inanition. Underdeveloped enzymes and metabolic pathways may decrease the infant's tolerance to excessive or inadequately balanced nutrient intakes. The amino acid requirements during rapid growth in infancy has been shown to be different from the amino acid pattern which merely prevents net nitrogen loss in adults (4).

The fluid turnover of the neonate during parenteral feeding amounts to about 14% of the body weight, while the corresponding figure for adults if 3 - 5%. To these figures must be added abnormal losses and their replacement. The high sensitivity of newborns to excessive sodium loads is only too well known (5,6). Furthermore, the kidney of the infant does not protect against inappropriately restricted fluid intake or excessive solute load as efficiently as the adult kidney does. Thus many normal infants can not concentrate their urine in excess of about 600 mOsm/1 (7) while the normal adult kidney can attain up to 1400 mOsm/1. Another difference between adults and infants is the fact that the skeleton of the neonatal patient continues to grow and hence to accumulate phosphorus and calcium, while the adult patient confined to bed tends to demineralize the skeleton.

In general, intravenous feeding programs for newborns have to be composed within fairly narrow limits taking into account both normal requirements and the particular patient's ability to utilize the nutrients. Even the normal infant has a very limited capability to compensate for errors in the infusion program.

Methods

Parenteral feeding program. Our routine and experimental parenteral feeding programs have all been composed to imitate human milk fairly closely. Table I compares our present routine program with the contents of human milk. Normal intake of milk has been set to 130 ml per kg per day, which is a reasonable figure for the second week of life (3).

Table I: Composition of human milk and parenteral feeding program

	Units	Contents of 130 ml human milk	Absorption as percentage of intake	Parenteral feeding program: amounts/kg/ 24 H.
Calories	g	98		106
Nitrogen	"	0.43	84	0.46
Fat	"	5.4	92	4.0
Carbohydrate	"	9.6		13.5
K^+	mmoles	2.3	83	3.3
Mg^{++}	"	0.09	63	0.3
Phosphate	"	0.64	89	1.5
Ca^{++}	"	0.85	55	1.0
Na^+	"	2.4	91	2.3

Table II describes our routine program based on 1-Aminofusin 5% "Pfrimmer" and 20% Intralipid "Vitrum". The feeding program is increased stepwise during the first week of life, the levels of intake specified in Table II being attained between the 4th and 6th day of life. Thus adaptive enzyme systems should have enough time to reach sufficient activity to deal with full nutritional intakes.

Table II: Parenteral feeding program for newborns and infants based on L-Aminofusin 5% "Pfrimmer" and Intralipid "Vitrum"

Preparation	Amount Per Kg Per 24 Hours	
L-Aminofusin 5% "Pfrimmer"	60 ml (456 mg N)	36 kcal
15% Glucose	50 ml	30 "
The above solutions are mixed and supplemented with:		
KH_2PO_4 (1 mmole/ml)	1.5 ml	
Ca-"Sandoz" (10%, 0.25 mmole/ml)	4 ml	
Pancebrin "Lilly"	0.15-0.20 ml	
20% Intralipid "Vitrum"	20 ml (4 g fat)	40 "

1 mg Vitamin K is given every 3 days.

We can see no reason to use the method of parenteral "hyperalimentaion" (8), i.e., to increase the intravenous supplies significantly since normal growth can be obtained without such "hyperalimentation."

Infusion methods. The liberal use of the isotonic soy bean oil emulsion Intralipid ("Vitrum") in our parenteral feeding program reduces the resultant hypertonicity of the solutions. Intralipid and the solution containing amino acids, carbohydrates, electrolytes, etc., are infused simultaneously, the infusion sets being connected by a Y-piece. In most cases, such infusions can be made through peripheral veins (scalp and extremity veins) for several weeks. It is necessary to change the infusion site every 1 - 3 days. The high complication rate which accompanies infusions into the vena cava is thus avoided. The use of broad spectrum antibiotics is also restricted to counteract the development of fungus sepsis.

The use of infusion pumps is strongly recommended. Without such pumps it is very difficult and time-consuming for the nursing staff to watch and adjust the drip rates. We use a syringe pump to inject about 60 ml 20% Intralipid in 24 hours, while a peristaltic pump (Braun Infusomat) delivers the 250 - 500 ml per 24 hours of the crystalline solution.

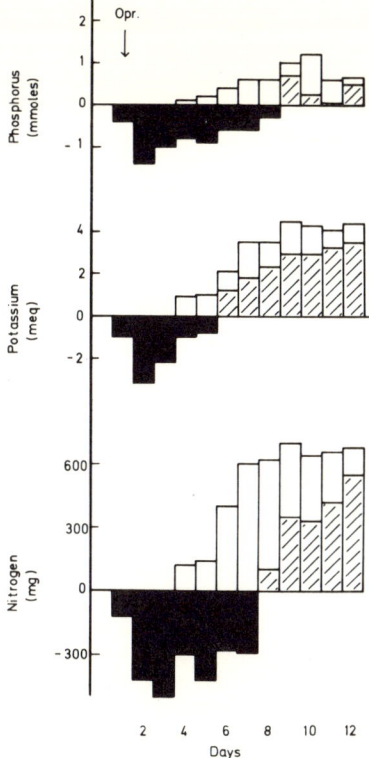

Fig. 1. Postoperative nitrogen, potassium, and phosphorus balances. Patient operated on 5th day of life for duodenal stenosis. No postoperative intakes whatsoever until the 3rd day, when 90 ml 5% glucose was infused. From the 4th postoperative day on increasing amounts of human milk was given. Cow's milk mixture was added from the 8th day. Balance diagrams are conventional, with negative balances marked in solid black.

Adequate postoperative parenteral feeding. A recent example of our postoperative balance recordings is shown in figures 2, 3, and 4. The positive balances, even in the immediate postoperative period, contrast sharply with the data in figure 1.

It may be of interest to point out a peculiar source of error in the nitrogen balance which is illustrated in the first seven postoperative days. The apparent nitrogen retention in this period was strikingly low, while the high, stable retention of potassium, magnesium and phosphorus indicated a normal growth rate and maturation of cellular tissue. It is known that the total red blood cell mass of the infant decreased sharply in his first week of life. Thus the serum bilirubin attained 16 mg%, and thereafter fell slowly. He also becamse anemic. The nitrogen from the catabolized hemaglobin is of course excreted in the urine, resulting in a low apparent nitrogen retention in spite of a high rate of nitrogen incorporation in newly synthesized protein. This phenomenon of low apparent nitrogen retention in spite of high retention of potassium, magnesium and phosphorus has also been observed in other of our patients with high postoperative bilirubin levels.

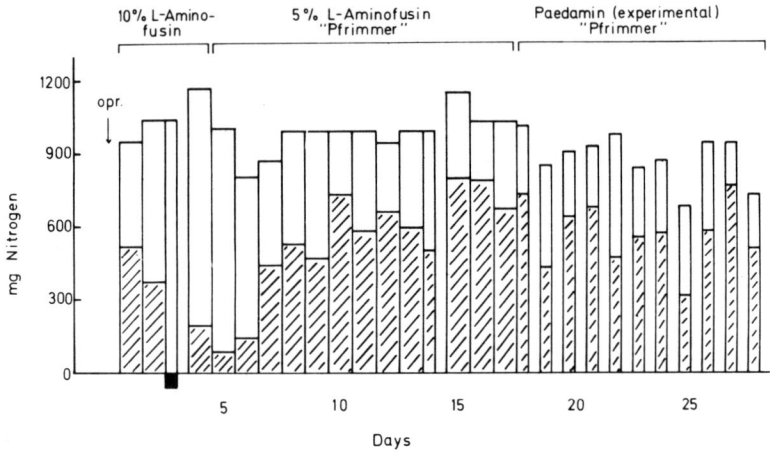

Fig. 2. Postoperative nitrogen balance on total parenteral feeding. Male patient (S.J.) operated on 3rd day of life for 11 Intestinal Atresias. Postoperative Hyperbilirubinemia and anemia. The parenteral feeding program described in Table II was used from the 5th to the 17th postoperative day. Very similar program the first 4 days. From the 18th day on, the experimental Paedamin "Pfrimmer" was used. Conventional balance diagrams with positive balances marked with oblique lines. Narrow columns indicate interruption of infusions lasting more than 2 hours during the 24 h. period.

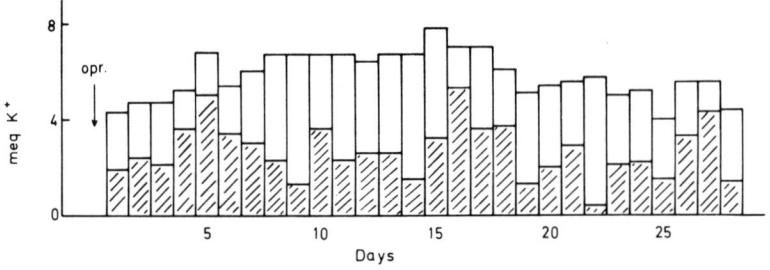

Fig. 3. Postoperative potassium balance. Same patient as Fig. 2.

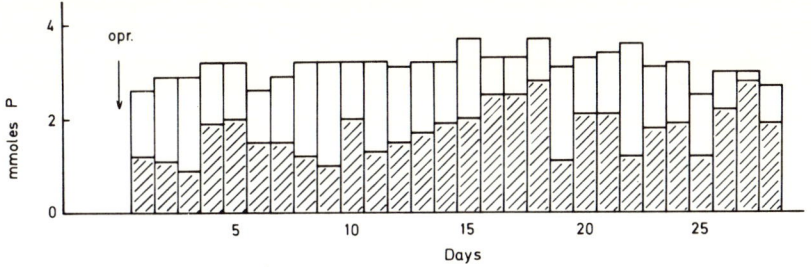

Fig. 4. Postoperative phosphorus balance. Same patient as Fig. 2.

The relative retentions of several intracellular elements are further analyzed in Table III. The retention ratios pertaining to the infusion program based on the commercial Aminofusin (Pfrimmer) (Tables I and II) were calculated from data obtained between the 7th to the 17th postoperative day. The first 7 postoperative days had to be excluded from these calculations, due to the influence of the above source of error in the nitrogen balance.

Table III: Nitrogen balance and element retention ratios

Element or Element ratio	Breast feeding	Bottle feeding	Solution A	Solution B
Nitrogen mg N/kg/24 h.	213± 48	304± 32	276 (61% of intake)	226 (64% of intake)
K/N meq/g	5.6	6.7	4.4	4.2
Mg/N meq/g	0.6	1.8	1.4	0.4
P/N mg/g	81	168	84	99
Ca/P meq/mmol	1.38	2.4	2.3	2.0

Solution A: Parenteral feeding based on L-Amino-fusion 5% "Pfrimmer" and Intralipid 20% "Vitrum". S.J. 7.-17.day.
Solution B: Parenteral feeding based on Paedamin "Pfrimmer" and Intralipid 20% "Vitrum". S.J. 18.-28.day.

Effects of parenteral feeding: Evaluation criteria. Statistical assessment of survival rates with and without adequate parenteral feeding regimes is difficult to obtain. Probably one has to resort to clinical impressions, experience and judgment. Weight gain during parenteral feeding and eventual brain size (head circumference) are important criteria, but the latter does not appear to have been used up to now. Increase of body weight may reflect alterations of body water and salts alone. Furthermore, weight gain due to accumulation of fat tissue, which appears to be almost proportional to caloric intake, may occur in the absence of normal maturation and formation of the tissues. We regard balance studies as indispensable in the analysis of weight gain in terms of synthesis of normal cells and maturation of tissues. Nitrogen balances alone are not sufficient, since the accumulation of normal tissue can be assumed only if normal proportional retentions of other intracellular elements like potassium, magnesium and phosphorus can also be documented. Thus the metabolic evaluation of our feeding methods is based mainly on multiple simultaneous balance measurements.

Absence of clinical complications and abnormal biochemical patterns is a desirable attribute to an ideal parenteral feeding method. Thus the pattern of serum amino acids during infusions should remain within the normal ranges. Specifically, concentration patterns known or supposed to elicit brain damage should be avoided. Furthermore, monitoring of the serum triglyceride level during constant-rate infusions is also required: If the infused triglycerides are metabolized like normal chylomicra, the serum triglyceride level will not exceed that seen during and after oral feeding.

Commercial intravenous amino acid preparations: Shortcomings in pediatric use. The amino acid patterns of commercial mixtures of crystalline l-amino acids are not based upon the requirement patterns which have been established for infants (12) Furthermore, the Aminofusin-series contains a surplus of non-essential nitrogen, which represents an unnecessary nitrogen load, and which limits the percentage of the administered nitrogen which can be incorporated into newly synthesized tissue protein. Another commercial solution, Vamin (Vitrum) is, like human milk protein, deficient in non-essential nitrogen (13), a fact which may be of advantage in uremic patients, but which may not be conducive to optimal utilization of the infused nitrogen in the normal patient. Our limited experience with this preparation confirms this view.

The pattern of essential amino acids in solutions for intravenous use has up to now been based on amino acid requirements during oral intake. It must be appreciated, however, that intravenously administered amino acids are not initially "filtered" by the liver. The optimal pattern of amino acids may therefore be different in parenteral feeding.

The commercial amino acid preparations do not contain carbohydrates and amino acids in the right proportions needed in infant nutrition. Mixing of amino acid solutions and carbohydrate preparations prior to use is therefore necessary. The electrolyte patterns are also far from ideal, necessitating further additions. The registered preparations do not, of course, contain phosphorus or calcium in significant amounts, since they are primarily designed for use in adult patients.

The extensive mixing of carefully measured volumes which is thus necessary prior to infusions is time-consuming and very susceptible to errors and contaminations.

Paedamin (Pfrimmer): Experimental amino acid/carbohydrate/electrolyte solution for pediatric use. The Pfrimmer factory has kindly supplied an experimental amino acid/carbohydrate/electrolyte solution the composition of which corresponds closely to the amounts specified in Table I. The pattern of essential amino acids takes into account the results of Snyderman et al. (12). The E/T ratio (ratio between essential and total amino acids) is 2.8 which is close to the composition of the cellular proteins which are accumulated during growth.

The experimental solution is easy to use, since the mixing prior to infusion is greatly simplified: Amino acid-electrolyte solution and carbohydrate solution (bottles with accurate volume supplied by "Pfrimmer") are mixed through a special needle before the infusion starts. A suitable multi-vitamin preparation is then added.

Results: Balance data, serum amino acid patterns and triglyceride levels

Postoperative parenteral undernutrition. Figure 1 displays balance data which are representative of the period preceding the introduction of adequate parenteral feeding programs for postoperative newborns (2). The case illustrated in figure 1 is a favorable one, since oral feeding was possible as early as the fourth postoperative day. Yet, the postoperative net loss of nitrogen, potassium and phosphorus is considerable. The over-emphasis on the ultimate complete recovery of patients like these can lead to difficulties in the appraisal of the advantages of adequate parenteral feeding.

The switch to the experimental Paedamin-based infusion program in the third period shown in Figures 2 to 4, increased the average nitrogen retention to about 64% of the infused amount. This occurred in spite of the frequent interruptions and technical difficulties with the infusions during this period. The retention ratios in the last column of Table III confirm that the electrolyte composition of Paedamin (Pfrimmer) is close to the optimal pattern.

Table IV: Serum Amino Acids (Micromoles per Litre)

	Blood drawn during Intravenous feeding (Patient G.E., Age 9 months)		Blood drawn after Oral feeding	
	Amino acid source		1 hour after meal (fish, milk, potatoes) Patient G.E.	Means 4 hours after meals Snyderman et al.(14)
	L-Aminofusin "Pfrimmer" x)	Paedamin "Pfrimmer" x)		
Glutamic Acid	350	183	132	
Glutamine	65	204	545	
Alanine	320	250	208	275
Proline	386	160	303	200
Valine	137	220	248	194
Glycine	435	190	162	164
Threonine	61	93	99	144
Serine	97	111	121	114
Leucine	84	131	133	104
Lysine	180	155	166	102
Tyrosine	39	39	51	82
Ornithine	60	47	50	71
Histidine	52	61	65	62
Isoleucine	42	88	75	59
Phenylalanine	78	66	78	52
Arginine			84	47
Aspargine	11	18	29	31
Taurine	37	48	59	24
Citrulline				21
Methionine	44	17	24	21
Butyrine	16	14	47	19
Asp. Acid				3
Cystine				27

x) Nitrogen intake was 325 and 290 mg per kg per 24 h with L-Amino fusin and Paedamin respectively.

Serum amino acid patterns during infusion. Snyderman et al. have studied the relationship between oral feeding and serum amino acid patterns in infants (14). Their data provides a background against which serum aminograms during intravenous infusions can be evaluated (Table IV, last column). However, the blood samples analyzed by Snyderman et al. were drawn four hours after meals, and may thus represent a semi-fasting state rather than the absorptive phase. Moreover, the relationship between serum amino acids and the composition of the administered protein or amino acid mixture was complex since the level of daily protein intake had considerable influence. The figures in the last column of Table IV refer to protein intakes of 3 - 3.5 g per kg per 24 h.

Table IV lists three different sets of plasma amino acid values obtained in a single patient, a 9-month old male infant, during infusion of two different amino acid mixtures as well as one hour after a mixed oral meal. Differences between the two aminograms obtained during infusions were in the directions which could be expected on the basis of differences in the amino acid composition of the two infused mixtures. This pertained to essential as well as semi-essential and non-essential amino acids. No evidence for toxic elevated concentrations of individual amino acids was obtained. For cystine reliable figures could not be given. The figures for phenylalanine and tyrosine are of special interest, since tyrosine is very slightly soluble and has to be partly or completely replaced by phenylalanine in intravenous solutions. The corresponding expected tendency for the phenylalanine/tyrosine ratio to be elevated during intravenous infusions proved not to be particularly striking. L-Aminofusin supplies three times as much glycine as Paedamin. The corresponding change of serum glycine level during infusion was by a factor of 2.3. The data for proline was similar.

Serum triglyceride levels during infusions of Intralipid "Vitrum". Five assays of total serum triglyceride level in our patient S.J. were performed between the 5th and 26th postoperative day. The figures varied between 113 and 221 mg per 100 ml. Daily measurements of serum triglycerides in other patients on total parenteral feeding in the first postoperative week have given similar results. Provided the patient is not subject to excessive stress, the total triglyceride level usually stays between 100 and 200 mg per 100 ml during constant-rate Intralipid infusions (4 g fat per kg per 24 h). According to Hallberg et al. (15) the rate of lipolysis increases with increasing serum Intralipid concentration up to a level corresponding to about 170 mg per 100 ml total serum triglycerides (endogenous lipoproteins plus exogenous Intralipid). If total serum triglycerides exceed this value during Intralipid infusion, the organism cannot increase the liplytic rate further to cope with the increasing Intralipid load. Infusions of Heparin (500 I.U. per 24 h) have not given any consistent reduction in the steady state serum triglyceride level. Apparently some degree of adaptation to the Intralipid infusions does occur, since the serum triglyceride levels tend to fall in the first week of intravenous feeding in spite of increasing daily doses of Intralipid.

Conclusions

Our general conclusion based on clinical experience and balance recordings during the last seven years (9, 10, 11) is that normal rate and pattern of growth can easily be obtained even in the immediate postoperative period in neonates by means of adequate intravenous feeding provided the clinical state of the patient is good enough to support normal cellular metabolism. True "hyperalimentation" appears unnecessary since about 100 kcal and about 2.5 g amino acids per kg per 24 hous have proved to be sufficient except in small prematures.

The use of the isotonic soy bean oil emulsion Intralipid "Vitrum" to cover about 40% of the caloric requirement permits peripheral vein infusions to be used exclusively for up to several weeks. Furthermore, this regimen ensures a sufficient supply of the essential linoleic acid. Linoleic acid deficiency which is inevitable unless at least 10% of the caloric needs is covered with Intralipid, leads to skin changes, alterations of cellular and other membranes, and increases the caloric requirement by about 20% (16).

Our knowledge of serum amino acid patterns during infusions is still insufficient to draw definite conclusions. However, our data are in accordance with the general view that the essential and semi-essential amino acids should be supplied in a requirement adapted pattern, while the non-essential amino acids should be provided in such proportional amounts as not to distort the serum amino acid pattern during infusion. The interconversion of different non-essential amino acids appears not to be sufficiently rapid to abolish the effects of the infused amino acid mixture on the serum amino acid pattern during infusion.

REFERENCES

1. Davies, P.A. and Davies, J.P. Lancet, 1216-1219, 1970.

2. Knutrud, O. The water and electrolyte metabolism in the newborn child after major surgery. Universitetsforlaget, Oslo, Norway, 1956.

3. Slater, J.E. Brit. J. Nutr., 15, 83-97, 1961.

4. Irwin, M.I. and Hegsted, D.M. J. Nutrition, 101, 539-566, 1971.

5. Finberg, L., Kiley, J., and Luttrell, C.N. JAMA, 184:187, 1963.

6. Rostad, R., Blystad, W. and Knutrud, O. Clinical Pediatrics, 3:1-4, 1964.

7. Ziegler, E.E. and Fomon, S.J. The Journal of Pediatrics, 78: 561-568, 1971.

8. Wilmore, D.W., Groff, D.B., Bishop, H.C. and Dudrick, S.S. J. Ped. Surgery, 4:181-189, 1969.

9. Borresen, H.C. and Knutrud, O. Acta Paediat. Scand., 58:420-421, 1969.

10. Borresen, H.C., Coran, A.G. and Knutrud, O. Annals of Surg., 172: 291-301, 1970.

11. Borresen, H.C. Nutr. Metabol., 13, 1972. In press.

12. Snyderman, S.E. et al. Reviewed in ref. 4.

13. Snyderman, S.E., Holt, L.E., Jr., Dancis, J., Roitman, E., Boyer, A. and Balis, M.E. Nutr., 78:57-62, 1962.

14. Snyderman, S.E., Holt, L.E., Jr., Norton, P.M., Roitman, E., and Phansalkar, S.V. Pediat. Res., 2:131-144, 1968.

15. Hallberg, D., Schubert, O., and Wretlind, A. Nutritio et Dieta, 8:245-281, 1966.

16. Houtsmuller, U.M.T. Evaluation of Modern Foods as Sources of Lipids, in Lipids, Malnutrition, and the Developing Brain, a Ciba Foundation Symposium. Associated Scientific Publishers, Amsterdam, 1972.

CONTROLLED PARENTERAL NUTRITION OF PREMATURE INFANTS

P. Jürgens, D. Dolif, C. Panteliadis and C. Hofert

Medical Department, General Hospital St. Georg and the
Children's Hospital Borgfelde, Hamburg, Federal Republic
of Germany.

The first successful parenteral feeding of a human based on
sound theoretical and experimental principles was carried out by
Abderhalden et al. (1909) in a 9-year old child. As early as
1939, several investigators (Farr et al., 1939; Shohl et al., 1939)
infused casein hydrolysates intravenously into children and obtained
positive nitrogen balances. There is currently an extensive pre-
and postoperative clinical experience concerning complete parenteral
nutrition of newborns suffering from severe anomalies of the intestinal tract and severely ill infants in whom for various reasons
a sufficient oral nutrition is not possible (Børresen et al., 1970;
Dudrick et al., 1970; Dudrick, 1971; Erdmann, 1970; Farr and
MacFadyen, 1939; Filler et al., 1969; Heine and Kirchmair, 1963;
Hofert et al., 1971; Kirchmair and Heine, 1967; Schmidt, 1966;
Wilmore et al., 1969). Major operations of the intestinal tract
in neonates were made possible only because of parenteral nutrition
(Dudrick et al., 1970; Dudrick, 1971; Wilmore et al., 1969).
Certainly it appears that parenteral nutrition will provide an
important advance in the clinical management of a variety of new-
born problems associated with caloric restriction.

Despite the variety of clinical experiences it must not be
neglected that exact requirements of the individual nutrients as
well as their limiting values for children have not yet been determined under the conditions of parenteral nutrition. The metabolic effects of intravenous infusion of differently composed
amino acid solutions in premature infants should be evaluated in
controlled experiments. Such nutritional studies should aim at
obtaining data on amino requirements and regulatory mechanisms of
amino acid metabolism for this period of life.

The problems of the oral feeding of immature prematures, especially during the first days and weeks of life, are generally known. Normally the caloric needs of the premature infant would have been delivered via the placenta. Thus for premature infants artificial parenteral nutrition may be an especially appropriate form of food intake.

We have investigated parenteral nutrition in fifteen male premature infants with an average body weight of 1670 gm (1,000 to 2,400 gm) without congenital organic or enzymatic defects. The infants were nourished exclusively by the parenteral route for seven to thirteen days (average 10 days) (Table 1).

Table 1: Dates of the Male Prematures

	Test Series I	Test Series II
Number of Cases	10	5
Body Weight in gm	1620 (1000-2050)	1770 (1365-2400)

The first 10 premature infants (Test Series I) were given the synthetic L-amino acid solution using an infusion which had provided the most favorable nitrogen balances and constant serum concentrations of amino acids in adults (Bansi et al., 1967; Coats, 1967; Dolif and Jürgens, 1971; Heller, 1967; Jürgens et al., 1970). The five prematures in the second Test Series received an amino acid solution modified according to the results of the first Test Series.

Table 2: Infusion program of 15 male prematures (ml solution per kg body weight per day).

Solution	Test Series I	Test Series II
Aminofusin L 600	60	
Aminofusin L ANN		65
Lipofundin S 10%	25	9
Solution of 10% glucose and 10% fructose	77	75
Solution of 13.6% KH_2PO_4	1.2	1.2

Parenteral nutrition was always started within the first 24 hours of life. All infusion solutions (Table 2) were administered simultaneously through a venous catheter in an umbilical scalp vein. The carbohydrate and amino acid solutions were infused continuously using infusion pumps.

The children were maintained in incubators in a special intensive care unit for premature infants. Before starting parenteral nutrition, we determined body weight, blood pH, total nitrogen in the stool as well as total nitrogen, urea nitrogen, α-amino nitrogen and creatine in the 24-hour urine. These measurements were carried out daily during and for several days after termination of parenteral nutrition. Measurements of serum urea nitrogen, electrolytes and the free amino acids were carried out on the first day of life at intervals during and after the period of infusion.

Parenteral nutrition did not cause any evident local or general complications. Three premature infants of the first Test Series died between the fourth and sixth day of life; however, none of their metabolic parameters differed from those found in the surviving children. The cause of death in two of the infants was extensive amniotic aspiration; one child had severe hyaline membrane disease.

The infants tolerated an intravenous fluid intake of 164 and 150 ml respectively (Table 3) without complication and no evidence for excessive water retention. Weight changes were in accordance with the nitrogen changes measured.

Table 3

Nutrient		Recommended Oral Intake	Amounts given per kg per day Series I	Series II
Water	(ml)	150	164	150
Sodium	(mEq)	2	2.2	2.2
Potassium	(mEq)	1.5	3.0	2.7
Magnesium	(mEq)	2.5	0.3	0.3
Calcium	(mEq)	5.4	0	0
Chloride	(mEq)	4.2	2.7	2.7
Phosphate	(mEq)	21	9.5	5.7
Acetate	(mEq)		2.2	2.2
Malate	(mEq)		1.4	1.4
Protein	(gm)	2.5	3.0	2.8
Calories	(kcal)	115	90	70

Opinions concerning optimal electrolyte requirements in prematures are contradictory (1970; Hartmann, 1968; O'Brien et al., 1954). Several authors (Hartmann, 1968; Rickham, 1957) do not recommend infusion of electrolytes during the first days of life or suggest the infusion of only small amounts of NaCl. In our experience, serum electrolytes remain in the normal range when 2.2 mEq of sodium, 2.7 to 3.0 mEq of K, 0.3 mEq of Mg and 2.7 mEq of Cl per kg body weight are infused daily. The administration of electrolytes should always depend on the actual serum values and the calculated losses.

Administration of 5.7 mEq of phosphate per kg body weight per day to test group 2 resulted in a positive phosphate balance with a serum value of 4.5 mEq per kg body weight per day. If calcium was not provided simultaneously, the daily calcium losses amounted to 0.1 mEq per kg body weight per day in this test series. During long term parenteral nutrition, administration of phosphorus, calcium, trace elements and vitamins is an absolute requirement (Dudrick et al., 1970; Schmidt, 1966).

The tolerance of prematures to relatively high amounts of fluid permits that 100-120 kcal per kg body weight can be given parenterally (Børresen et al., 1970; Dudrick et al., 1970; Erdmann, 1970; Wilmore et al., 1969). This caloric intake corresponds to about twice the basal requirements for premature infants and should support growth (Reardon, 1959). During a seven to thirteen day period of complete parenteral nutrition, prematures in Test Series I received 90 kcal per kg per day and those in Test Series II received 70 kcal per kg body weight daily (73-83% of total calories as carbohydrates and sugar alcohols). In both test series, about 5% of the total caloric intake was provided as amino acids. Thus the calories provided parenterally to the infants differ from mother's milk and more closely correspond to the intake of adults.

The removal of polyols, sorbitol and xylitol from the blood of newborns does not differ from that of adults (Bässler et al., 1962; Toussaint, 1969). Therefore, these polyols can also be used with amino acid solutions for the parenteral nutrition of premature infants (Griem and Lang, 1962; Lang, 1963; Mehnert, 1966; Prellwitz and Bässler, 1963).

There has been considerable experience with parenteral administration of fat emulsions to young babies (Melichar, 1967; Schmidt, 1962; Schmidt, 1966; Schmidt, 1967) and these have been tolerated without major side effects. Daily administration of 2.4 g to a maximum of 5 g of fat per kg body weight is recommended. Both of our test series received intravenously lipid, 2.5 g per kg in Series I and 0.9 g per kg in Series II. This was given as Lipofundin S (10% soybean oil emulsion, Braun Melsungen). No complications were observed. Fat emulsions are not only an indispensable caloric source in parenteral nutrition, but also provide essential fatty acids (Coats, 1969: Hansen, 1963). Fat emulsions clearly have a protein-sparing effect in infants (Schmidt, 1962; Schmidt, 1966; Schmidt, 1967).

If one considers the protein requirement of the mature newborn, 2 - 2.5 gm per kg body weight (Hegstedt, 1964) and also considers the change in protein requirement during the first weeks of life (Fig. 1), the daily nitrogen intake of 0.5 gm and 0.44 gm per kg body weight in this and other series (Borresen, 1970; Dudrick et al., 1970; Erdmann, 1970; Schmidt, 1962; Schmidt, 1966; Schmidt, 1967; Wilmore, 1969) appears to be the lower range of protein required.

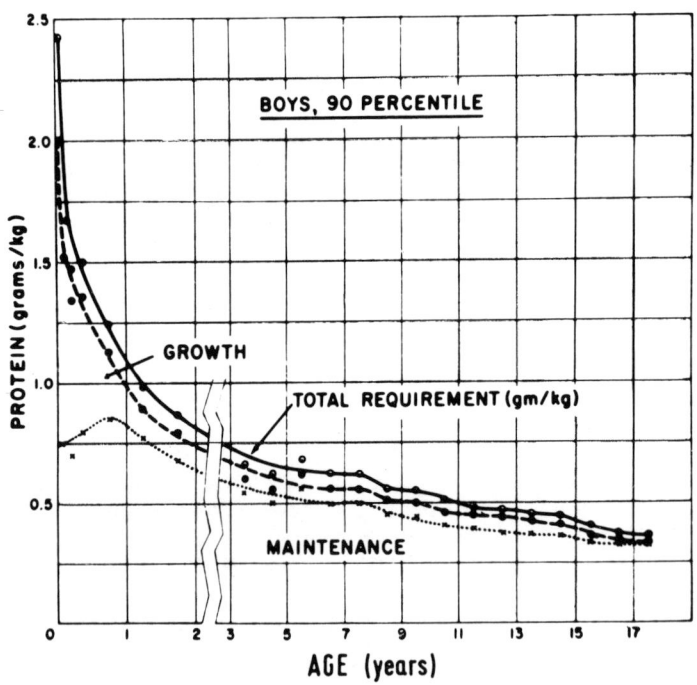

Fig. 1. Protein requirements during childhood (Hegsted, 1964).

The L-amino acid solution used in Test Series I contained all essential amino acids in proportions of the FAO reference protein, and in amounts considered safe by Rose (1957). In addition, the solution contained histidine 0.1 gm %, arginine 0.3 gm %, alanine 0.51 gm %, glutamic acid 0.83 gm %, glycine 0.6 gm % and proline 0.7 gm % (Table 4).

Table 4: Amino acid requirements and daily intake during parenteral nutrition.

Amino Acid mg/kg	Minimal Requirements of Infants*	Test Series I	Test Series II
Isoleucine	126	115	103
Leucine	150	131	111
Lysine	103	132	119
Methionine	45+	139	71
Cystine			8.5
Phenylalanine	90±	153	93
Tyrosine			20
Threonine	87	70	75
Tryptophan	22	32	32
Valine	105	101	78
Histidine	34	63	38
Arginine		220	212
Alanine		410	368
Aspartic Acid			191
Glutamic Acid		895	429
Glycine		390	372
Proline		447	224
E/T – Ratio	~ 0.7	1.7	1.6

* Taken from Holt et al., 1960.
+ In presence of adequate cystine
± In presence of adequate tyrosine

The L-amino acid solution used in the second Test Series contained essential amino acids and histidine as recommended by Holt et al.(1960. The solution also contained aspartic acid 0.29 gm %, cystine 0.01 gm %, tyrosine 0.03 gm %, glycine 0.57 gm %, proline 0.35 gm % and alanine 0.37 gm %. Thus, both solutions contained the amino acids described as essential by Balestrieri et al. (1968). All essential amino acids were administered at least in amounts meeting the minimum values advised by Holt et al. (1960; Snyderman et al., 1962; Snyderman et al., 1959). The ratio of essential to total amino acids (E/T ratio), amounting to 1.7 or 1.6, was nearly identical in the amino acid solutions used.

Children in both Test Series remained in positive nitrogen balances during the test period. Despite the different amino acid compositions of the solutions used, the mean daily nitrogen retentions achieved were 0.3 gm per kg body weight and 0.27 gm per kg body weight in Test Series I and II respectively. In both Test Series, 61% of the nitrogen supplied was retained. The total nitrogen excretion averaged 0.2 gm per kg body weight per day in Test Series I and was only slightly higher than the endogenous daily nitrogen excretion. In Test Series II nitrogen excretion was 0.17 gm per kg body weight per day, also within the normal range for endogenous daily nitrogen excretion (Hegstedt, 1964). In both Test Series, the urinary nitrogen was almost exclusively urea nitrogen. In Test Series I, α-amino nitrogen excretion in the urine was in the upper normal range, approximately 20 mg per day (Schmidt, 1966; Schreier et al., 1957). In Test Series II, α-amino nitrogen could not be detected in the urine. In both Test Series, the total nitrogen content of the stool averaged 10 mg per kg body weight per day.

Clearance values for most amino acids in premature infants are higher than those in adults (Greestein and Winitz, 1961; Schreier et al., 1957). Thus storage of free amino acids is unlikely in the premature infant. In both Test Series, the serum urea values were in the upper normal range (Fig. 2).

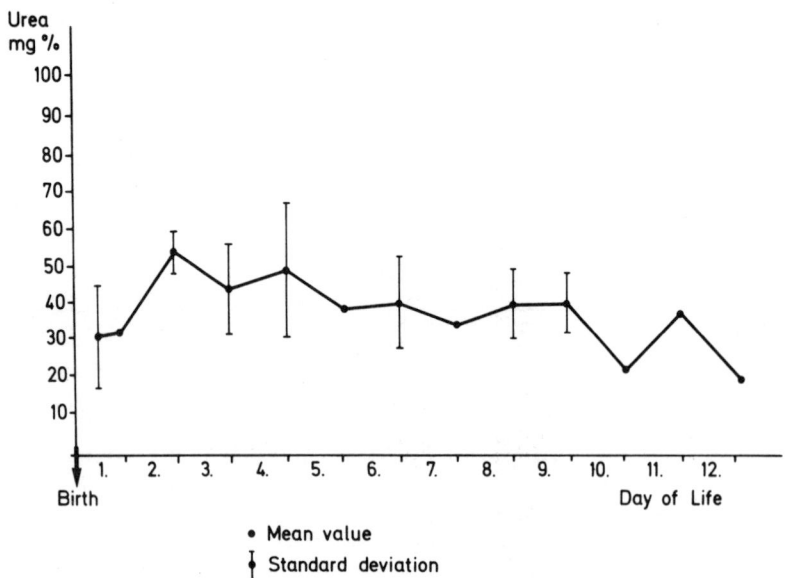

Fig. 2. Urea concentration in the serum of 15 male prematures under the conditions of complete parenteral nutrition.

Thus it appears that in both Test Series the nitrogen retained was needed for protein synthesis. From the mean daily nitrogen retention of 0.3 gm per kg body weight achieved during parenteral nutrition in Test Series I, one can calculate a daily weight gain of approximately 11 gm per kg body weight (0.3 x 6 x 6.25). The daily weight gain of 8 gm per kg body weight in Test Series I approximates the theoretical value. The estimates for water re-

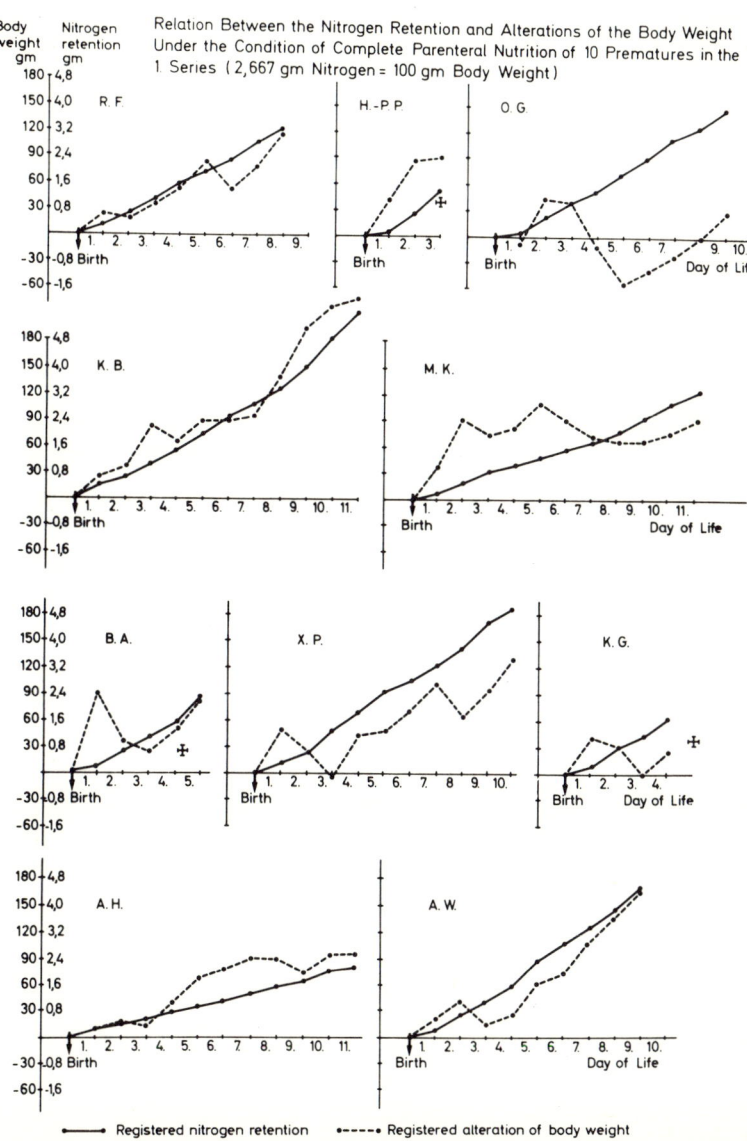

Fig. 3

tention, approximately 11 ml per kg body weight, is also in good accordance with these values.

The relationship between nitrogen retention and change of body weight in Test Series I is shown in Figure 3.

In Test Series II, the calculated theoretical daily weight gain was 10 gm per body weight given the daily nitrogen retention of 0.27 gm per kg body weight. However, the measured daily weight gain measured in this Test Series was only 3 gm per kg body weight. Analysis of the individual cases (Fig. 4) shows strikingly different nitrogen retention and weight changes in children J. Sch., M.B. and O. St., especially between the 2nd and 5th day of life.

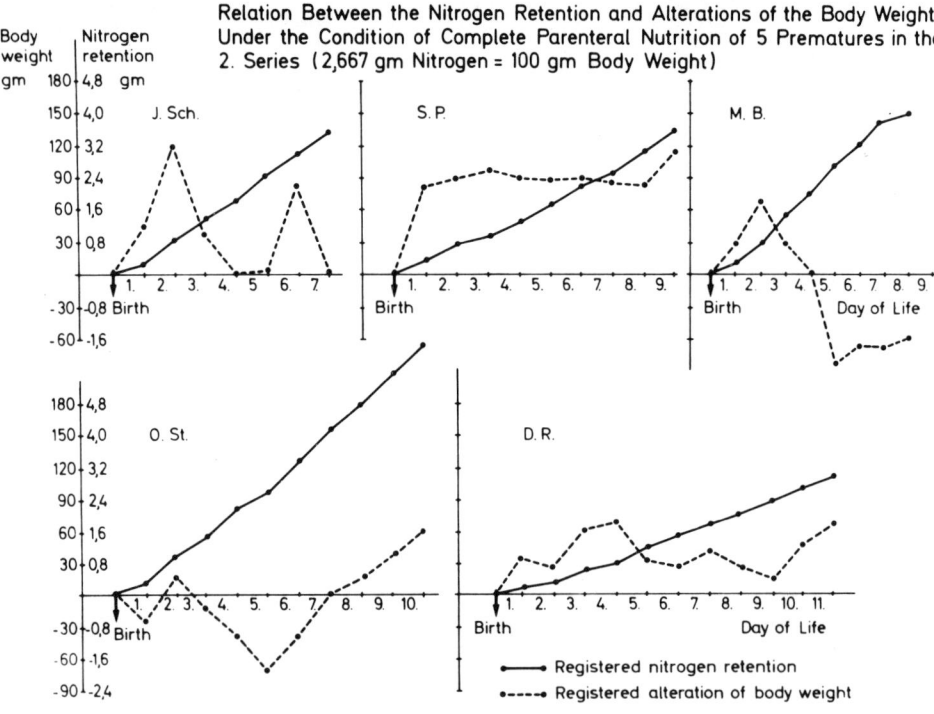

Fig. 4

These three children as well as two children in the first Test Series (O.G., X.P.) showed marked edema after birth. Infant J. Sch. received furosemid on two occasions. At the end of the infusion period, all children showed a complete regression of the edema. As 1 gm of nitrogen is equal to about 37 gm of tissue, there is an average excess of water amounting to 30 ml per kg body weight before starting parenteral nutrition for the infants in the first Test Series, and an excess of approximately 110 ml per kg body weight (Fig. 5) for the children in the second Test Series.

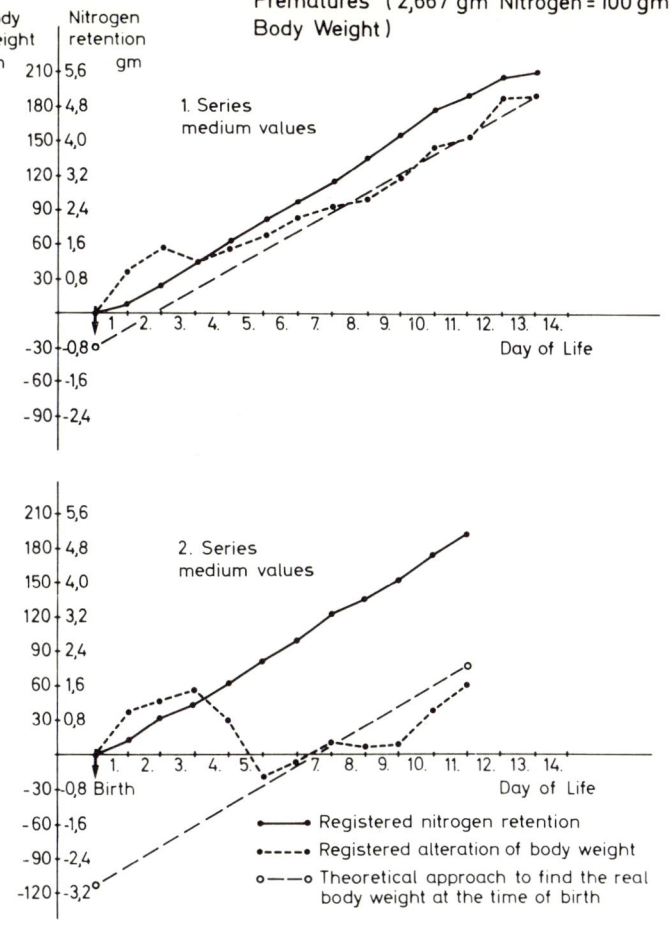

Fig. 5

Summarizing this first phase of our experimental data, we have to assume that both Test groups did equally well despite the different amino acid solutions administered. Urine losses of free amino acids were significant when the first amino acid solution was infused. However, this did not exceed normal values. These results agree with earlier reports (Farr and MacFadyen, 1939; Shohl, 1943; Shohl and Blackfan, 1940; Shohl et al., 1939) showing positive nitrogen balances with infusion of such different solutions as casein hydrolysate and synthetic amino acid solutions composed according to the requirements of the rat. The growing human organism has such large requirements for nitrogenous nutrients that the ideal amino acid composition of parenteral solutions cannot be established by nitrogen balance alone.

Further information on the amino acid metabolism can be obtained by column chromatography determination of the free amino acids in serum (Holt et al., 1960; Longenecker, 1965; Longenecker and Hause, 1959; Longenecker and Hause, 1961; Snyderman et al., 1962; Snyderman et al., 1968; Snyderman et al., 1959). There are, however, two major difficulties. Normal amino acid values for the pre-term fetus are not known. Also, all previous studies of serum amino acid concentration in prematures have been carried out on infants receiving discontinuous nutrition. Obviously this would not be the case if the premature infant had remained in utero. With the above pitfalls in mind, we have used the normal values established by Snyderman et al., (1968) obtained during oral nutrition of young mature babies one hour after intake of a formula-milk diet containing 3 - 3.5 gm of protein per kg body weight. Our hypothesis is that proper amino acid balance in a parenteral solution is achieved when under the conditions of infusion the amino acid concentrations in the serum are within the normal range reported by Snyderman.

During the period of parenteral nutrition, we measured serum amino acid concentrations in each child at intervals of 2 - 4 days. The measurements were made in such a way that values obtained from several children were obtained during the same study period. However, since the serum concentrations of some amino acids varied in different children, concentrations of some amino acids appeared to fluctuate.

The fasting serum concentration of valine, 6 - 18 hours after birth, was distinctly below and that of lysine was greater than that found after oral nutrition was begun. The basal concentrations of these amino acids also differed from values reported by Snyderman et al. (1968) during oral nutrition of young infants. The marked increase of the lysine-valine ratio (Kraut and Simmermann-Telchow, 1962) with feeding may reflect the prevailing protein catabolism at that time. In prematures, adequate nutrition should therefore be started during the first hours of life.

When the amino acid solution adapted to the amino acid requirements of adults was given, we observed that children receiving 139 mg of methionine per kg per day and 153 mg of phenylalanine per kg per day exhibited a considerable increase of the serum concentration of methionine (Chase et al., 1968; Komrower and Robins, 1969) (Fig. 6), and a variable increase in phenylalanine levels until the second day of therapy. Methionine and phenylalanine concentrations became normal in all children until the end of the eighth day of infusion with the same amino acid mixture. This was probably due to the activation of adaptive enzymatic processes

Concentration of Methionine, Cystine, Phenylalanine and Tyrosine in the Serum of Prematures Under the Condition of Complete Parenteral Nutrition

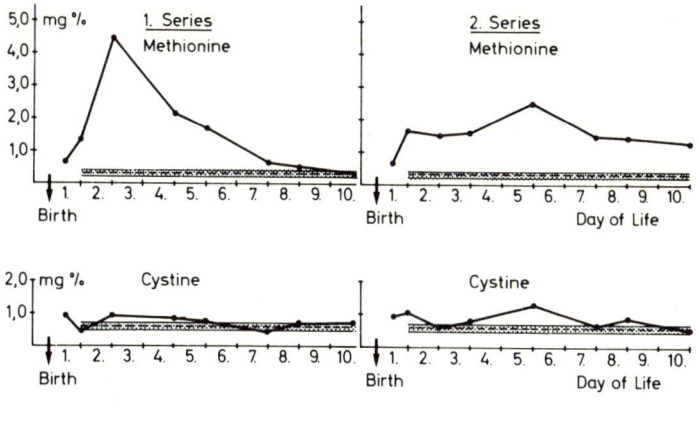

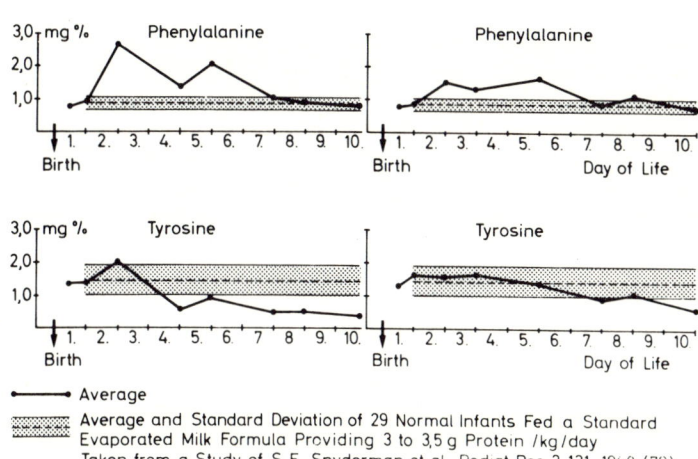

Fig. 6

favoring metabolism of these amino acids (Chase et al., 1968; Greenstein and Winitz, 1961; Komrower and Robins, 1969; Lang, 1970). At the same time, serum concentrations of cystine remained constant in the upper normal range although no cystine was given. By contrast serum concentrations of tyrosine dropped below the normal range after the second day of life. This is possibly a consequence of immaturity of tyrosine-oxydase-system during this period of life.

In the second Test Series, the supply of methionine was decreased to 71 mg per kg and that of phenylalanine to 93 mg per kg. Simultaneously, 8.5 mg of cystine per kg and 20 mg of tyrosine per kg were added to the amino acid solution. This resulted in a lesser rise in serum methionine concentration than observed in the first Test Series. At the same time, the rapid induction of counter-regulatory processes, observed before, failed to appear so that the serum methionine concentrations in the second Test Series were still increased at the end of the tenth day of infusion. In the second Test Series, the serum cystine concentrations were slightly higher than those of the first Test Series. A further reduction of methionine supply is planned. Cystine itself does not seem to be necessary. When the supply of phenylalanine was reduced, the serum concentrations of this amino acid showed a smaller increase in the second Test Series as compared with the first Test Series. In the second Test Series, the serum tyrosine concentrations were significantly higher than those seen in the first Test Series. By the 5th day of infusion, however, they decreased to values considered to be in the lower normal range or less. Thus, an increase in the supply of tyrosine seems to be necessary. In view of these changes in serum concentration, tyrosine may be an essential amino acid during this period of life.

Despite the similarity of amounts administered (Table 4), the serum concentrations of threonine, valine, isoleucine, leucine and lysine were clearly different in Test Series I and II (Figure 7). Amino acid imbalances and/or amino acid antagonisms have to be assumed to be the reason (Greenstein and Winitz, 1961; Harper, 1964; Harper, 1968; Lang, 1970) for this variation. In Test Series II, the serum concentrations of valine and leucine were roughly in the normal range; the concentrations of threonine, isoleucine and lysine were clearly above normal. Thus further corrections of the amounts of these amino acids seem to be indicated.

Although the histidine concentrations in Test Series I were higher than the normal values if 63 mg per kg body weight was supplied (Figure 8), we found that in Test Series II the serum histidine concentrations were in the normal range when 38 mg per kg were given. In both Test Series, the serum concentrations of arginine and ornithine were for the most part above the normal values, when 212 - 220 mg of arginine per kg body weight were given. At the same time, serum concentrations of citrulline were rather low in both Test Series. An optimal supply of the amino acids of the Krebs-Henseleit-cycle can only be found by further detailed experiments.

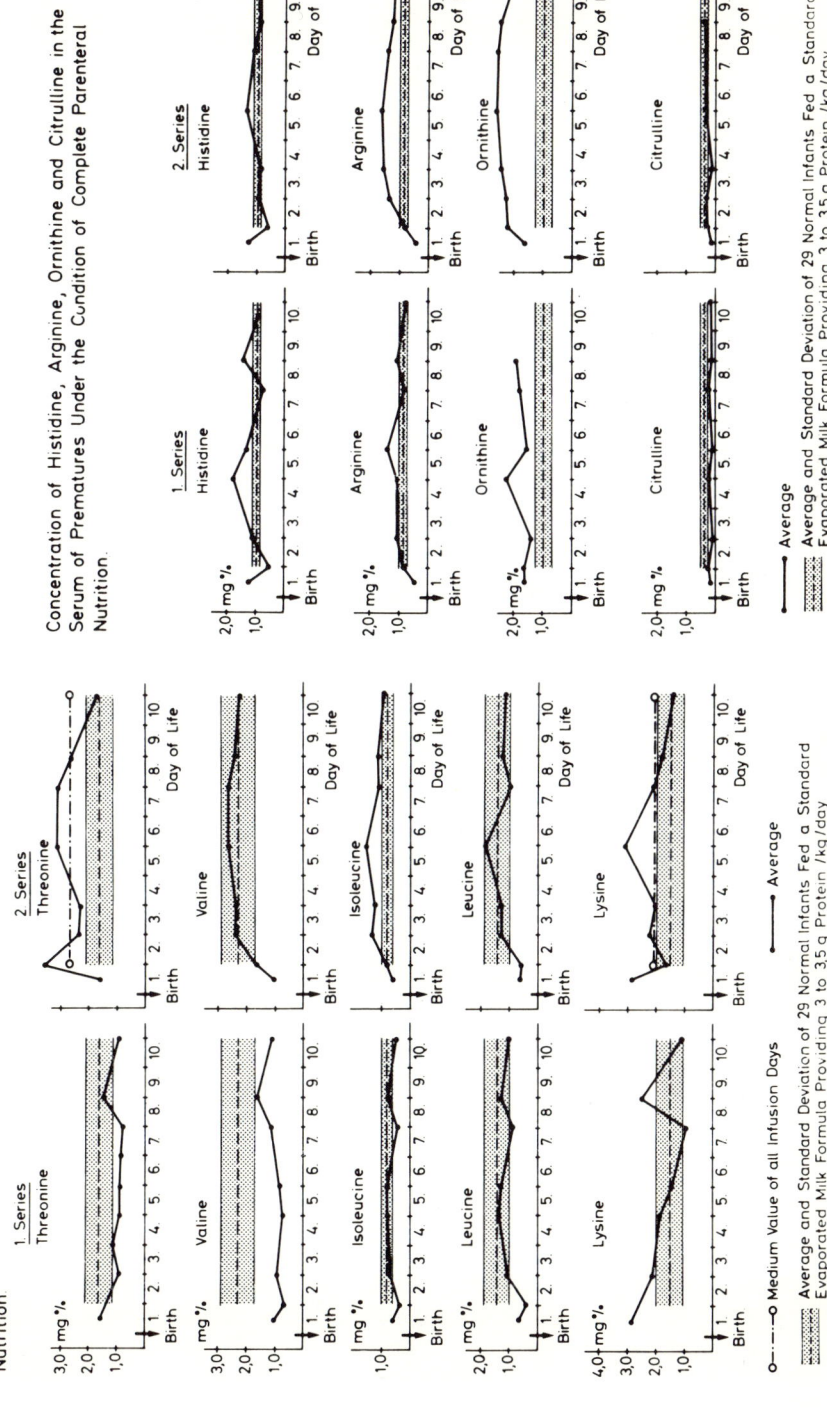

Fig. 7

Fig. 8

In our previous own studies on parenteral nutrition of adults, we had recognized proline as being an essential amino acid (Dolif and Jürgens, 1970; 1971; Jürgens et al., 1968, 1970). This finding has been confirmed by Balestrieri et al. (1968) who directly measured the proline synthesis rate in adults.

It is rather certain that proline is also essential for infants. All amino acid solutions which are used in pediatrics should therefore contain proline. In our studies on parenteral nutrition in adults, a strong relation between the requirements of the classical essential amino acids and proline was established. In case of deficient or decreased supply of dicarbonic acids, approximately 5 gm of proline per 1 gm of threonine had to be given in order to achieve a constant fasting concentration of proline and optimal nitrogen balances (Dolif and Jürgens, 1970; 1971; Jürgens et al., 1970; Jürgens and Dolif, 1968). The infusion of such an amino acid solution in the first Test Series caused an unexpectedly high increase in the serum proline concentrations in prematures (Figure 9). This apparently different metabolic response may possibly be explained by relating the supply of proline to the body weight: in the above cited experiments on parenteral nutrition, adults had received about 100 mg of proline per kg body weight; prematures, however, received about 450 mg per kg body weight. In the second Test Series, proline was therefore reduced to 224 mg per kg body weight. In this Test Series, the serum proline concentrations were only slightly higher than the normal values. It was surprising that the hydroxyproline concentrations in the first Test Series were not at all influenced by the modification of the serum proline concentration, despite a significantly increased hydroxyproline excretion in the urine (12.2 mg per day). Because of the methodical difficulties of separating glutamine and asparagine by column chromatography and the possibility of conversion of these biological amines the chemical preparation of the serum and the chromatography itself, the total of the measured values: aspartic acid + glutamic acid + asparagine + glutamine is indicated. There are no reference values available for assessing these figures. Since in both Test Series about the same total values have been obtained in spite of different supplies (Test Series I = 895 mg of glutamic acid per kg body weight, Test Series II = 429 mg of glutamic acid per kg body weight and 191 mg of aspartic acid per kg body weight), it may be assumed that the respective supplies were within the physiological regulatory system.

The distinct increase of the serum glycine concentrations after glycine administration of 390 mg per kg body weight in Test Series I and 372 mg per kg body weight in Test Series II (Figure 10) is not surprising. In previous studies on parenteral nutrition in adults, we were able to prove that limitation of glycine turnover to a maximum of about 200 mg per kg body weight per day

CONTROLLED PARENTERAL NUTRITION OF PREMATURE INFANTS 193

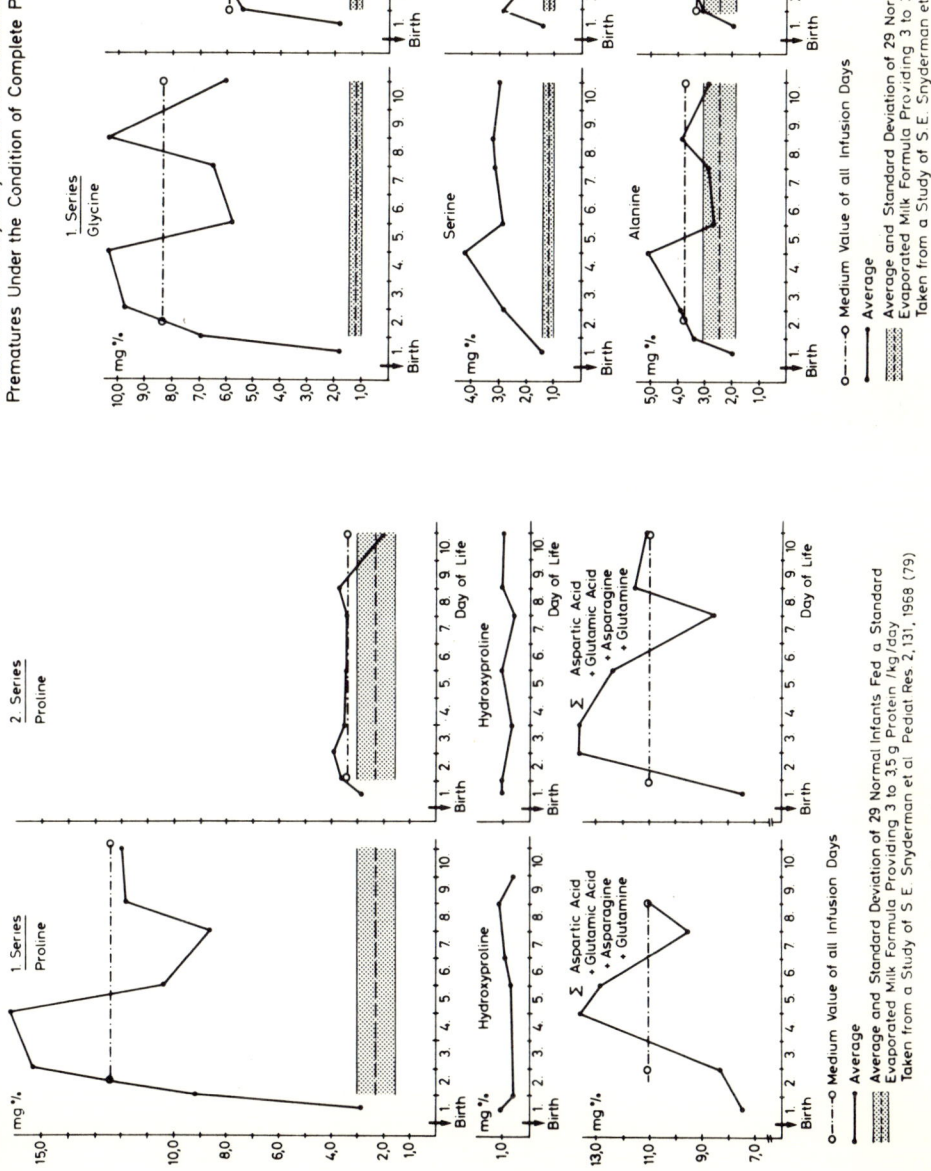

Fig. 9

Fig. 10

was independent of other factors of protein metabolism (Dolif and Jürgens, 1970; 1971; 1971; Jürgens et al., 1970; Jürgens and Dolif; 1970). This value seems to us to be a useful guide for further artificial feeding of prematures.

As has also been described in previous studies on parenteral nutrition of adults (Dolif and Jürgens, 1970; 1971; 1971; Jürgens et al., 1970; Jürgens and Dolif; 1970), the serum serine concentrations were markedly increased in parallel with serum glycine concentrations in the premature infants of both test series when high levels of glycine but not serine were given. The metabolic conversion of glycine into serine obviously takes place very rapidly in prematures so that an additional serine supply is undesirable, at least when glycine is given in amounts close to maximal glycine turnover.

Despite a reduced supply, the serum alanine concentrations were generally above the normal values in Test Series II. The metabolic effects of a further reduced alanine supply are to be studied.

Although parenteral infusions of amino acid solutions have been used successfully in pediatrics for 33 years, the precise requirements of amino acids and the physiological limits for artificial feeding of very young children have not yet been established. Our studies of these problems are by no means completed. A number of methodical problems have been pointed out. For these reasons it has been impossible to provide final data on all amino acids requirements for prematures. In many cases, only approximate values can be given. What is the importance of these measured values? These studies show very clearly that it is not possible to draw any conclusions concerning the metabolic effects of a nutritive amino acid solution used in infants merely from nitrogen balance studies. The rate of protein synthesis rate during the period of artificial feeding might conceal possible disturbances of the synthesis of important structures or functions of certain organs, which for example are well known to us as secondary damage of an inborn error of the amino acid metabolism.

This study has shown that large changes in serum amino acid concentrations may arise when infusions of amino acid solutions are given to premature infants.

REFERENCES

Abderhalden, E., Frank, F. and Schittenhelm, A. Hoppe Seylers Z. physiol. Chem., 63, 215, 1909.

Bassler, K.H., Unbehaun, V. and Prellwitz, W. Biochem. Z., 336, 35, 1962.

Balestrieri, C., Cittadini, D. and Giordano, C. Life Sci., 7, 1033, 1968.

Bansi, H.W., Dolif, D. and Jürgens, P. in Fortschritte der parenteralen Ernährung. ed. N. Henning u. G. Berg. Pallas. Lochham bei Munchen, 1967.

Børressen, H.C., Coran, A.G. and Knutrud, O. in Fortschritte der parenteralen Ernährung. ed. G. Berg. Georg Thieme, Stuttgart, 1970.

Chase, H.P., Volpe, J.J. and Laster, L. J. Clin. Invest., 47, 2099, 1968.

Coats, D. in Fortschritte der parenteralen Ernährung. ed. N. Henning and G. Berg. Lochham bei Munchen, 1967.

Coats, D. Z. Ernährungswiss, 9, 401, 1969.

Dolif, D. and Jürgens, P. in Fortschritte der parenteralen Ernährung, ed. G. Berg. Georg Thieme, Stuttgart, 1970.

Dolif, D. and Jürgens, P. Z. Ernährungswiss, suppl., 10, 24, 1971.

Dolif, D. and Jürgens, P. in Bilanzierte Ernährung in der Therapie. ed. K. Lang, W. Fekl and G. Berg, Gerog Thieme, Stuttgart, 1971.

Dudrick, S.J., Long, J.M., Steiger, E. and Rhoads, J.E. Med. Clin. North America, 54, 577, 1970.

Dudrick, S.J. in Bilanzierte Ernährung in der Therapie. ed. K. Lang, W. Fekl and G. Berg. Gerog Thieme, Stuttgart, 1971.

Empfehlungen zur parenteralen Ernährung. (Cooperation of numerous authors). Med. u. Ernähr., 11, 201, 1970.

Erdmann, G. in Fortschritte der parenteralen Ernährung. ed. G. Berg. Georg Thieme, Stuttgart, 1970.

Farr, L.E. and MacFadyen, D.H. Proc. Soc. Exp. Biol. (N.Y.),
    42, 444, 1939.

Filler, R.M., Eraklis, A.J., Rubin, V.G. and Das, J.B. New Eng.
    J. Med., 281, 589, 1969.

Greenstein, J.P. and Winitz, M. Chemistry of Amino Acids.
    John Wiley and Sons, New York & London, 1961.

Griem, W. and Lang, K. Wschr., 40, 801, 1962.

Hansen, A.E., Wiese, R.F., Boelsche, A.N., Hoggard, M.E., Adam,
    D.-H.D. and Davis, H. Pediatrics, 31, 171, 1963.

Harper, A.E. in Mammalian protein metabolism. ed. H.N. Munro and
    J.B. Allison, Vol. II. Academic Press, New York and London,
    1964.

Harper, A.E. Amer. J. Clin. Nutr., 21, 358, 1968.

Hartmann, G. Mschr. Kinderheilk., 116, 578, 1968.

Hegstedt, D.M. in Mammalian protein metabolism. ed. H.N. Munro
    and J.B. Allison, Vol. II. Academic Press, New York and
    London, 1964.

Heine, W. and Kirchmair, H. Z. Kinderheilk., 88, 186, 1963.

Heller, L. in Fortschritte der parenteralen Ernährung. Pallas,
    Lochham bei München, 1967.

Hofert, C., Dolif, D., Jürgens, P. and Panteliadis, C. Symposion
    über Therapie lebensbedrohlicher Zustände bei Säuglingen und
    Kleinkindern, 8 and 9.10. 1971, Mainz (in press).

Holt, L.E., Gyorgy, P., Pratt, E.L., Snyderman, S.E. and Wallace,
    W.M. Protein and Amino Acid Requirements in Early Life,
    University Press, New York, 1960.

Jürgens, P., Bansi, H.W. and Dolif, D. in Parenteral Nutrition,
    ed. H.C. Meng and D.H. Law. C.C. Thomas, Springfield,
    Illnois, 1970.

Kirchmair, H. and Heine, W. in Fortschritte der parenteralen
    Ernährung. ed. N. Henning and G. Berg. Pallas,
    Lochham bei Munchen, 1967.

Komrower, G.M. and Robbins, A.J. Arch. Dis. Childhood, 44, 418,
    1969.

Kraut, H. and Zimmermann-Telchow, H. Nutr. et Dieta (Basel), 4, 22, 1962.

Lang, K. Med. u. Ernähr., 4, 45, 1963.

Lang, K. Biochemie der Ernährung. 2nd edition, Steinkopff, Darmstadt, 1970.

Longenecker, J.B. in Newer Methods of Nutritional Biochemistry, ed. A.A. Albanese, Vol. I. Academic Press, New York and London, 1965.

Longenecker, J.B. and Hause, N.L. Arch. Biochem., 84, 46, 1959.

Longenecker, J.B. and Hause, N.L. Amer. J. Clin. Nutr., 9, 356, 1961.

Mehnert, H. in Parenterale Ernährung. ed. K. Lang, R. Frey and M. Halmagyi. Springer, Berlin-Heidelberg-New York, 1966.

Melichar, C.S. in Fortschritte der parenteralen Ernährung. ed. N. Henning and G. Berg. Pallas, Lochham bei München, 1967.

O'Brien, D., Hansen, J.D.L. and Smith, C.A. Pediatrics, 13, 126, 1954.

Prellwitz, W. and Bassler, K.H. Klin. Wschr., 41, 196, 1963.

Reardon, H.S. Pediatr. Clin. North America, 6, 181, 1959.

Rickham, P.P. The Metabolic Response to Neonatal Surgery. University Press, Harvard, Cambridge, Massachusetts, 1967.

Rose, W.C. Nutr. Abstr. Rev., 27, 631, 1957.

Schmidt, G.-W. Mschr. Kinderheilk., 110, 485, 1962.

Schmidt, G.-W. in Handbuch der Kinderheilkunde. Springer, Berlin-Heidelberg, New York, 1966.

Schmidt, G.-W. in Fortschritte der parenteralen Ernährung. ed. G. Berg. Pallas, Lochham bei Munchen, 1967.

Schreier, K., Ittensohn, R., Hans, U. and Sievers, W. Z. Kinderheilk., 79, 165, 1957.

Schreier, K. and Stieg, H. Z. Kinderheilk., 68, 563, 1950.

Shohl, A.T. J. Clin. Invest., 22, 257, 1943.

Shohl, A.T. and Blackfan, K.D. J. Nutr., 20, 305, 1940.

Shohl, A.T., Butler, A.M., Blackfan, K.D. and MacLachlan, F.
    J. Pediatrics, 15, 469, 1939.

Snyderman, S.E., Holt, L.E., Jr., Dancis, J., Roitman, E., Boyer,
    A. and Ballis, M.E. J. Nutr., 78, 57, 1962.

Snyderman, S.E., Holt, L.E., Jr., Norton, P.M., Roitman, E. and
    Phansalkar, S.V. Pediat. Res., 2, 131, 1968.

Snyderman, S.E., Prose, P.H. and Holt, L.E., Jr. J. Dis. Childhood,
    98, 459, 1959.

Toussaint, W. Mschr. Kinderheilk., 117, 262, 1969.

Wilmore, D.W., Groff, D.B., Bishop, H.C. and Dudrick, S.J.
    J. Pediat. Surgery, 4, 181, 1969.

TOTAL INTRAVENOUS ALIMENTATION IN LOW BIRTH WEIGHT PREMATURE INFANTS

William C. Heird, M.D., John M. Driscoll, Jr., M.D., Robert W. Winters, M.D.
Department of Pediatrics, Columbia University College of Physicians and Surgeons, and Babies' Hospital, Columbia-Presbyterian Medical Center, New York, New York

We have given total intravenous alimentation to nine infants weighing less than 1200 grams at birth. Table I shows the pertinent clinical data pertaining to these infants. There were three males and six females; birth weights ranged from 720 to 1150 grams. Gestational age, as determined by last menstrual period, neurological examination and neuroelectrical techniques, was between 26 and 30 weeks, with eight of the nine infants' weights being appropriate for gestational age. All infants were given intravenous nutrients as an attempt to increase the caloric intake above that usually achieved in infants of this weight. Associated clinical problems included pulmonary insufficiency of prematurity, recurrent apnea and necrotizing enterocolitis. No infant had respiratory distress syndrome.

Table I. Clinical Data Pertaining to Intravenously Alimented Infants

1. Number of Infants   9
    Male   3   Female   6
2. Birth Weight
    Range (gms.): 720-1150   Average (gms.): 890
3. Gestational Age (wks.)   26-30
4. Associated Clinical Problems
    Pulmonary Insufficiency   4
    Recurrent Apnea   5
    Necrotizing Enterocolitis   1

Table II. Data Pertaining to Technique of Intravenous Alimentation

Group	Number of Infants	Age	Duration of IV Alimentation	Daily Weight Gain (gms.)	
				Range	Average
I	6	less than 2 days	19-24 days	1.1-12.1	7.4
II	3	11-18 days	17-21 days	11.0-14.7	12.5

Table II summarizes some of the data pertaining to total intravenous alimentation. The infants can be separated into two groups, depending upon when intravenous alimentation was started. The first group consists of six infants who received the initial infusion within the first 48 hours of life. During the first 24 hours, the infusate contained 10 grams per kg of glucose and 2.5 grams per kg of beef fibrin hydrolysate as well as appropriate amounts of electrolytes, calcium, inorganic phosphate, magnesium, and vitamins. The solution was initially infused at 100 ml per kg per day, and progressively increased to 120-130 ml per kg per day. The glucose content was gradually increased to 25-27 gms per kg per day in accordance with the patient's tolerance for glucose, based on serial urine and blood sugar determinations. In this group of infants, total intravenous alimentation was continued for 5 to 24 days. Excluding one infant, who died after 5 days and who showed a weight loss of 16 grams per day, the average daily increase in body weight was 7.4 grams.

In the second group of infants, unsuccessful attempts at conventional feeding were made before total intravenous alimentation was started between the 11th and 18th day of life. Their infusions were continued for 17 to 21 days with an average weight gain of 12.5 grams per day.

The pattern of weight gain was similar to that seen in conventionally managed infants with weight loss initially, followed thereafter by a persistent weight gain. Figure 1 shows a comparison of weight curves for conventionally managed infants with those of our intravenously alimented infants. Those infants who received total intravenous alimentation within the first 48 hours of life had a shorter period of weight loss than expected and regained their birth weight sooner. A significant fraction of the time spent in regaining initial weight is undoubtedly due to an inability to provide adequate calories, defined as more than 100 cal per kg per day. Occurrence of hyperglycemia, requiring transient decreases in the glucose concentration of the infusate, prevented provision of adequate calories sooner.

TOTAL INTRAVENOUS ALIMENTATION IN PREMATURE INFANTS 201

PREMATURE GROWTH CHART

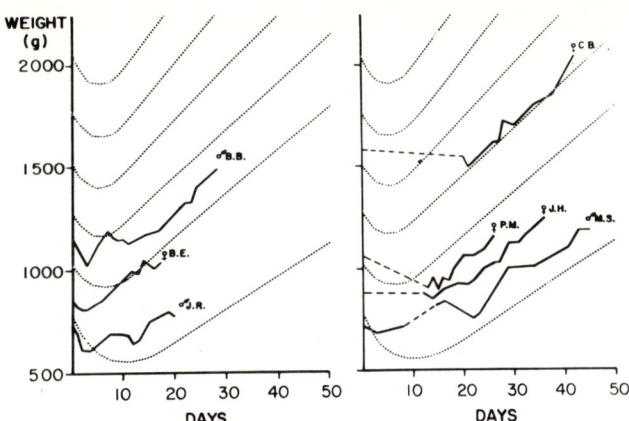

Fig. 1. Weight gains of intravenously alimented infants compared to that observed in conventionally managed infants. (Dancis, O'Connell, and Holt, 1948).

Figure 2 shows what appears to be a direct relationship between the time required to regain initial weight and the time required to achieve an intake of greater than 100 cal per kg per day.

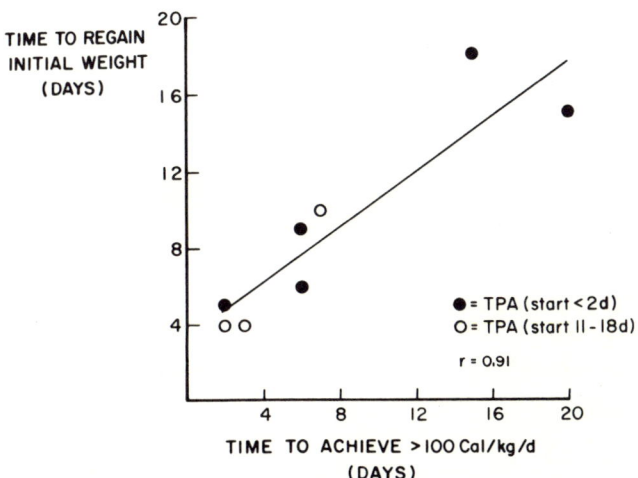

Fig. 2. Relationship between time required for intravenously alimented infants to achieve intake of 100 calories/kg/day and time required to regain birth weight.

Table III. Weight Gain and Nitrogen Balances Observed in Intravenously Alimented Infants Receiving Greater than 100 cal per kg per day.

	Initial Birth Weight	Duration of Intravenous Alimentation	Average Daily Weight Gain	Average Daily Nitrogen Balance
Average:	890 gms.	17.4 days	15.5 gms.	0.23 gms.
Range:	780-1180 gms.	14-20 days	11.0-19.3 gms.	0.20-0.28 gms.

Table III summarizes weight gain and nitrogen balance of five infants over periods of 14 to 20 days during which the caloric intake was at least 100 calories per kg per day. Over this interval, averaging 17.4 days, all infants showed a mean daily gain in body weight of 15.5 grams, and all showed positive nitrogen balances -- the average being 0.23 grams per day. The probable quality of the weight gain in these infants can be inferred from the nitrogen balance data in these five infants.

Table IV. Partition of Weight Gain of Intravenously Alimented Infants Receiving More than 100 calories per kg per day.

1.	Average Daily Weight Gain	15.5 gms.
2.	Average Duration of Intravenous Alimentation	17.4 days
3.	Total Weight Gain*	270 gms.
	a. Lean Body Mass**	120 gms.
	b. Other***	150 gms.

*   15.5 gms. per day X 17.4 days
** Total Nitrogen Balance X 17.4 days
*** Total weight gain <u>minus</u> Lean Body mass

As shown in Table IV, the average weight gain during the total period of intravenous alimentation was 270 grams. Assuming that deposition of one gram of nitrogen accounts for 30 grams of lean body mass, the total nitrogen retention over the 17.4 days -- i.e., 0.23 grams per day times 17.4 days -- accounts for 120 grams

of the total weight gain. This leaves 150 grams, or 15% of the
final weight, unaccounted for. It is unlikely that all of this
unexplained weight gain represents water. A cumulative gain of
150 grams of water would have produced massive edema; minimal,
if any edema, was detected. We conclude, therefore, that the
weight gain not due to an increase in lean body mass was due
largely, or entirely, to the deposition of fat.

Serial acid-base determinations performed during the period
of total intravenous alimentation showed a chronic respiratory
acidosis in almost all the infants. None of these acid-base
changes could be related to the infusate; rather, they were attributed to the pulmonary insufficiency of the very immature infant.

It has been suggested that hypercapnea and/or acidemia exert
a catabolic effect, making achievement of positive nitrogen balance impossible. To test this point, we compared the daily nitrogen balances of 5 patients with the daily values for blood pH
and plasma PCO_2. As shown in Figure 3, there was no clear relationship between the degree of either acidemia or hypercapnea,
and the degree of positive nitrogen balance achieved. Indeed,
positive nitrogen balances were achieved despite blood pH values
as low as 7.15 and PCO_2 values as high as 85 mm Hg.

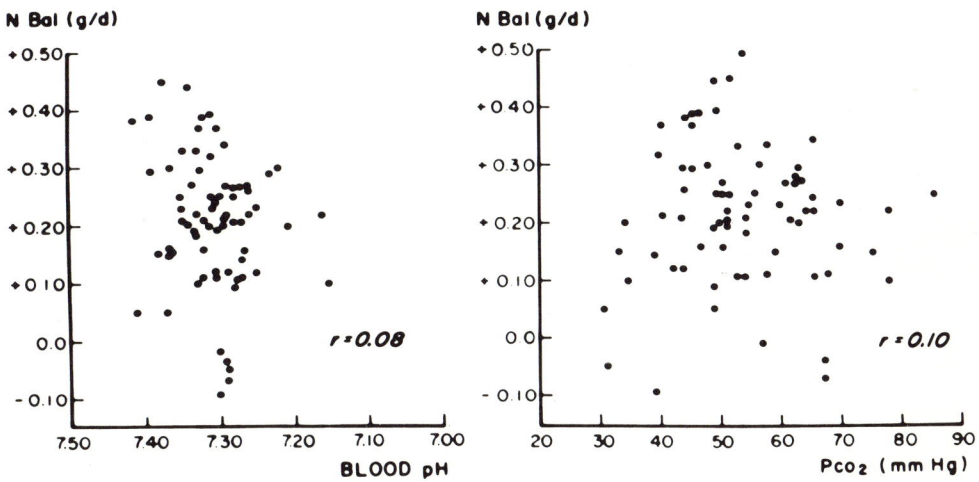

Fig. 3. Relationship between blood pH and plasma PCO_2 values to
degree of nitrogen balance observed in intravenously
alimented infants.

No significant deviations were observed in plasma sodium, potassium, calcium and phosphorus during the period of total intravenous alimentation. This fact is attributed to close chemical monitoring with appropriate daily adjustments of the electrolyte composition of the infusate being made on the basis of this monitoring.

Table V. Complications and Outcome of Intravenously Alimented Infants.

1. Metabolic Complications   11
    Hyperglycemia   8
    Azotemia   3

2. Deaths   4
    Candida Sepsis   1
    Unrelated Causes   3

3. Current Neurological Status of Survivors
    a. Normal   3 (12, 14, 16 mos. of age)
    b. Suspect   2 (14 and 15 mos. of age)

Table V shows the complications and outcome of the infants receiving total intravenous alimentation.

The only severe metabolic complication observed was hyperglycemia accompanied, on two occasions, by glycosuria, osmotic diuresis, and transient weight loss. Most such episodes occurred during the first 10 days of intravenous alimentation and were probably due to the infants' inability to handle the higher glucose concentrations. Three infants had blood urea nitrogen values between 40-50 mg% when the nitrogen source was administered at 4 grams per kg per day. Reduction of the nitrogen source to 2.5 grams per kg per day resulted in return of these values to normal limits.

One infant died of Candida Sepsis on the 5th day of infusion; no other catheter complications were encountered. Three other infants died from causes unrelated to the intravenous alimentation.

Of the five survivors, three infants were considered to be neurologically normal (at ages 16, 14 and 12 months); one was classified neurologically suspect (at age 15 months), and one was judged neurologically abnormal (at 13 months of age).

The findings of this study are highly preliminary. They do demonstrate that the intrinsic hazards of total intravenous alimentation in premature infants can be reduced to such a level as to justify further controlled clinical investigation of the technique. The results further show that even the very small premature infant can readily metabolize the infusate without serious variations in plasma electrolyte concentration, blood acid-base status, plasma calcium, and inorganic phosphate concentrations or blood glucose and urea nitrogen levels. The results do not prove, nor do they even suggest, that total intravenous alimentation has an accepted place in the routine management of the low birth weight infant.

INTRAUTERINE AMINO ACID FEEDING OF THE FETUS

Luz Heller

Department of Obstetrics and Gynecology of the University of Frankfurt/Main, West Germany

Treatment of placental insufficiency has been directed at improving the functional capacity of the placenta by means of hormonal substitution or by improving placental circulation. In many instances, however, the results have been unsatisfactory. We have therefore, considered the possibility of providing the fetus with nutrients independent of placental function.

Using labeled substances injected into the amniotic cavity, it has been demonstrated that the human fetus not only swallows amniotic fluid but also that the amount of amniotic fluid swallowed at term is as much as 220 to 750 ml per day (Gitlin et al., 1972; Plentl and Hutchinson, 1958; Plentl, 1958). This corresponds to 25 to 80% of the total amniotic fluid. Thus, fetal nutrition via the amniotic fluid should theoretically be possible.

During the intrauterine life protein deposition proceeds at an extremely high rate (O'Neill, 1971). Therefore, it is very important that adequate amino acid be provided to the fetus. On the other hand, there is a surprising paucity of exact information about amino acid metabolism in the fetus: No systematic study of the amino acid requirements of the fetus and also of the premature infant has been undertaken. It is possible, too, that certain amino acids not usually considered to be essential for the mature individual may be essential for the fetus and the premature infant because of limitations of certain enzyme activities (Snyderman, 1970).

The facts obtained so far may be summarized as follows:

1. The placenta is capable not only of transferring amino acids but also of concentrating them (Curet, 1970; Heller, 1971;

Young and Prenton, 1969). The mechanism by which the placenta maintains relatively normal concentrations of amino acids on the fetal side is unknown.

2. In the placenta free amino acids increase during pregnancy although they are already quite high by twenty weeks (Lorincz and Kuttner, 1969; Pearse and Sornson, 1969). Free amino acids are low in the infarcted placenta, and high in the toxemic placenta.

3. In the amniotic fluid there is a decreasing concentration of most amino acids with increasing gastational age. The maximum rate of decrease occurs between the twenty-second and the twenty-fourth week of gestation (O'Neill et al., 1971; Thomas et al., 1971).

In general however, there is only fragmentary information. Therefore, we have attempted to approach the problem of fetal amino acid requirements (Heller, 1971). Arterial and venous blood was obtained from the cord vessels of twenty two newborns at the time of birth before the cord was cut. Furthermore, the maternal internal iliac artery and vein were punctured during Cesarian section in five mothers in order to obtain blood from both sides of the uterine bed. Plasma amino acids were determined by column chromatography according to MOORE and STEIN with the Technicon apparatus. These results are shown in Table 1.

Table 1: Plasma amino acid concentrations in the arterial and venous blood of mother and newborn

Amino Acid	Mother (n=5)		Newborn (n=22)	
	arterial	venous	arterial	venous
Ornithine	0.73	0.68	1.86	1.67
Aspartic Acid	2.19	2.02	1.84	1.57
Threonine	2.81	2.23	3.68	3.55
Serine	2.20	1.91	2.61	2.30
Glutamine	9.92	8.92	10.89	10.74
Proline	2.24	1.81	2.26	2.10
Glycine	3.00	2.63	3.03	2.89
Alanine	5.09	3.89	4.20	4.04
Valine	2.14	2.13	3.13	3.03
Cystine	0.97	0.68	traces	-
Methionine	0.55	0.42	0.67	0.43
Isoleucine	0.69	0.62	1.01	0.94
Leucine	1.61	1.45	1.89	1.76
Tyrosine	0.93	0.77	1.27	1.17
Phenylalanine	1.68	1.46	1.72	1.45
Lysine	2.76	2.38	5.86	5.60
Histidine	1.68	1.55	2.03	2.01
Arginine	2.72	2.08	1.72	1.73
NH_3	4.97	2.28	4.12	3.99
A. A. without NH_3	41.96	33.82	49.60	47.03
A. A. with NH_3	46.92	36.10	53.72	51.01

Maternal values obtained from the internal iliac vein were similar to those obtained earlier from the cubital vein. The concentration of the individual amino acids in the arterial blood, however, was 10 to 20% higher than those of venous blood. This indicates that a considerable part of maternal plasma amino acids is taken up across the uterus. The total amino acid concentration of both arterial and venous blood of newborns was significantly higher than in the mother's blood. The venous amino acid concentration of the infant was also higher than those of maternal arterial samples indicating that the placenta concentrates amino acids against a gradient.

When the arterial amino acid pattern of the mother is compared with that of the child, we find that the individual amino acids in the placenta are concentrated very differently. In particular,

Table 2: Amino acid concentrations in the arterial plasma of mother and newborn

Amino Acid	Mother mg%	Newborn mg%	Values of Newborn in % of Values of Mother
Essential Amino Acids+			
Isoleucine	0.65	1.01	159
Leucine	1.61	1.89	118
Lysine	2.76	5.86	212
Methionine	0.55	0.67	123
Phenylalanine	1.68	1.71	102
Threonine	2.81	3.68	131
Valine	2.14	3.13	147
Semi-essential Amino Acids			
Alanine	5.09	4.20	82
Arginine	2.72	1.72	64
Histidine	1.68	2.03	119
Proline	2.24	2.26	100

+ Tryptophane not estimated.

the concentration of essential amino acids is much higher in the newborn than in the mother. Considering the maternal levels to be 100%, the values of infants were as follows: threonine 131%, valine 147%, isoleucine 159% and lysine 212%. The corresponding values for the semi-essential amino acids were much lower, and values for non-essential amino acids were slightly below 100% with the exception of glutamic acid. The placenta is therefore able to concentrate amino acids necessary for intrauterine development.

The amino acid pattern in the umbilical vein thus differs considerably from the amino acid pattern of the adult serum (Heller, 1971, 1972). Feeding premature infants amino acids in similar concentration proved to be very favorable with regard to the nitrogen balance

Table 3: Intrauterine amino acid feeding of the fetus.

Amino Acid Solution (5%)

Isoleucine	2.30
Leucine	2.50
Lysine	2.73
Methionine	1.45
Cystine	0.20
Phenylalanine	2.00
Tyrosine	0.50
Threonine	1.50
Tryptophane	0.45
Valine	1.90
Histidine	0.80
Arginine	4.20
Alanine	7.00
Aspartic Acid	3.00
Glutamic Acid	8.50
Proline	3.50
Glycine	8.00
Total Nitrogen	6.20
-Amino-Nitrogen	5.36

The nitrogen requirements of the fetus range between 0.5 and 0.7 g N/kg body weight daily.

For our studies, an amino acid solution was prepared which reflected what is made available to the fetus. The solution contained 6.2 gm N/l or 5.36 gm of -amino nitrogen. Since the nitrogen requirements of premature infants range between 0.5 and 0.7 gm N per kg body weight (Snyderman, 1970), the daily needs for nitrogen of a fetus weighing 2 kg can be covered by 200 ml of this solution. For our studies, 250 ml of this solution were slowly instilled into the amniotic fluid.

At specific intervals, amniotic fluid was aspirated for column chromatography. In addition, total nitrogen, -amino nitrogen, urea, uric acid, creatinine as well as certain enzymes were also determined.

The amino acids given intraamnially disappeared from the amniotic fluid at an unexpectedly rapid and rather uniform rate.

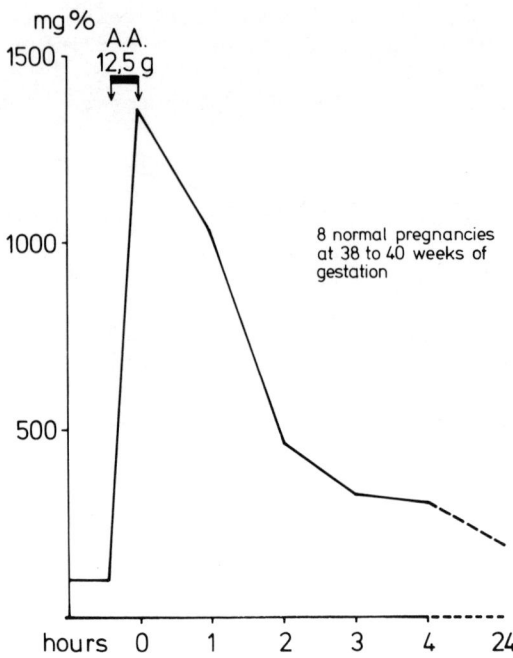

Fig. 1. Changes in the amniotic fluid concentration of amino acids given intra-amniotically.

Figure 1 shows the decreasing concentration of amino acids from the amniotic fluid as a function of time. After two hours, approximately two thirds of the amino acids given intraamnially disappeared; thereafter the rate of removal plateaued. After twenty four hours, the values returned to the baseline. The rapid disappearance of the amino acids from the amniotic fluid is difficult to explain. Even if one considers that the fetus swallowed the maximum amount of 50 ml per hour, the curve should have declined much slower. On the other hand, the permeability of the fetal membranes with regard to amino acids is extremely low as our studies have shown (Heller and Huter, 1970). The data are shown in Fig. 2.

For all amino acids, a mean permeability constant of $1.07 \pm 0.2 \times 10^{-3}$ cm/min results. On the basis of this value, a mean transfer rate of 800 mg per 24 hours may be calculated for these amino acids we measured (glycine, aspartic acid, threonine, serine, glutamic acid, proline, valine, cystine, methionine, isoleucine, leucine, tyrosine, phenylalanine, ornithine, lysine, histidine and arginine). If and to what extent amino acids are absorbed by the

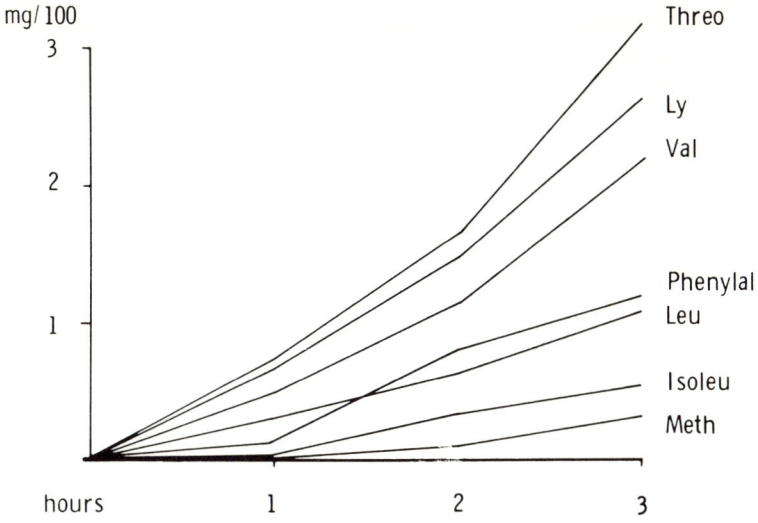

Fig. 2. Permeability of amino acids through amniotic membranes.

umbilical cord or the placenta is not known. Therfore, at the present time, it can only be stated that amino acids disappear very rapidly from the amniotic fluid, certainly much more rapidly than proteins which have a half life of about 30 hours in the amniotic fluid (Usatequi-Gomez et al., 1970).

Some remarkable observations could be made in cases of placental insufficiency. The diagnosis of a placental insufficiency was made by means of 4 parameters, at least 3 of them had to be positive (cardotocography, excretion of oestriol in the urine, measurement of the biparietal diameters by means of ultrasonics, human placental lactogen). In cases of placental insufficiency, amino acids were given intraamnially up to 12 times.

In four cases the reduced excretion of oestriol in the maternal urine returned to normal after intraamnial amino acid infusion. In one case, the silent monitor showed a normalization and the late decelerations in the oxytocin sensitivity test disappeared.

Despite the difficulty in interpreting the present findings, the intraamnial supply of amino acids seems to be a promising way to treat placental insufficiency.

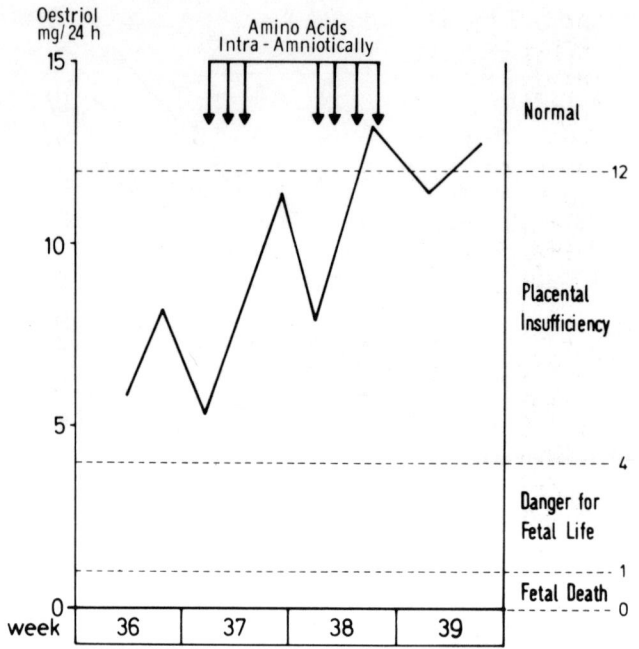

Fig. 3. Placental insufficiency. The effect of amino acids given intra-amniotically on maternal urinary oestriol excretion.

REFERENCES

Curet, L. B., Clin. Obstetr. Gynecol., 13, 586, 1970.

Gitlin, D., Kumate, J., Morales, C., Noriega, L. and Arenalo, N., Am. J. Obstetr. Gynecol., 113, 632, 1972.

Heller, L, Arch. Gynak., 211, 303, 1971.

Heller, L., in: A. W. Wilkinson, Proceedings of Parenteral Nutrition, Livingstone, Edinburgh, 1972.

Heller, L. and Huter, J., Rev. Espan. Obstetr. Ginecol., 29, 210, 1970.

Lorincz, A. B. and Kuttner, R. E., Am. J. Obstetr. Gynecol., 105, 925, 1969.

O'Neill, R., Morrow, G., Hammel, D., Auerbach, V. H. and Barness, L. A., Obstetr. Gynecol., 37, 550, 1971.

Pearse, W. H. and Sornson, H., Am. J. Obstetr. Gynecol., 105, 696, 1969.

Plentl, A. A. and Hutchinson, D. C., Proc. Soc. Exp. Biol., (N. Y.) 82, 681, 1953.

Plentl, A. A., in: C. A. Villee, Gestation, Macy, New York, 1958.

Snyderman, S. E., in: U. Stave, Physiology of the Perinatal Period, Meredith Corp., New York, p.441, 1970.

Thomas, G. H., Parmley, T. H., Stevenson, R. E. and Howell, R. R., Am. J. Obstetr. Gynecol., 111, 38, 1971.

Usatequi-Gomez, M., Hopkins, M. S. and De Castro, A. F., Obstetr. Gynecol., 36, 865, 1970.

Young, M. and Prenton, M. A., J. Obstet. Gynaec. Br. Commonw., 76, 333, 1969.

THE ROLE AND EFFECT OF PARENTERAL NUTRITION ON THE LIVER AND ITS USE IN CHRONIC INFLAMMATORY BOWEL DISEASE IN CHILDHOOD

Michael I. Cohen, Scott J. Boley, Fred Daum, Iris F. Litt and S. Kenneth Schonberg.

Division of Adolescent Medicine, Department of Pediatrics and Pediatric Surgical Service, Department of Surgery, Montefiore Hospital and Medical Center, and The Albert Einstein College of Medicine, Bronx, New York

THE LIVER

At the present time, specific and clear-cut clinical indications for the use of intravenous nutrition in the infant, child or adolescent with hepatobiliary disease do not exist. The indications for the implementation of such a life-support system on a temporary basis in children undergoing extensive operative procedures on the liver or biliary tract is similar to that found in any child undergoing radical surgery. The use of intravenous nutrition in such situations is, however, primarily if not solely, related to the need to nutritionally sustain such children during prolonged periods when they cannot tolerate oral feedings. There does not appear to be any specific advantage to this group of postoperative pediatric surgical patients as opposed to any other group of poorly nourished children following extensive surgery of the cardiovascular, neuromuscular or genito-urinary systems.

Those childhood metabolic disorders with a major component of hepatic inflammation or scarring, such as galactosemia, tyrosinemia, glycogenosis, and lipidosis may have very significant nutritional deficiencies as a result of the underlying pathophysiologic disturbance. Nonetheless, we have not been inclined to offer parenteral nutrition as the primary vehicle for nutritional rehabilitation in these patients. We have, instead, manipulated the oral feeding when feasible and given a regimen appropriate for the specific disease entity. Folkman and associates (1972) have recently utilized parenteral nutrition in the preoperative preparation of two children with glycogen storage disease prior to the performance of portacaval shunts. They describe marked diminution in the size of the liver as well as correction of many of the metabolic abnormalities. They were further impressed by the lack of postoperative metabolic problems previously seen after such shunts for glycogen storage disease (Boley, Cohen and Gliedman, 1970).

The following are several examples illustrating the application of intravenous nutrition as part of the post-operative regimen in youngsters with significant liver disease requiring extensive surgery. In L.V., a 1 year old female with extra hepatic biliary atresia, who developed large bowel obstruction secondary to cholestyramine resin treatment for intractable pruritis, intravenous nutrition was given for 14 days before the child died of candida sepsis (Cohen, Winslow and Boley, 1969). The second patient was W.B., a 14 year old male, who underwent subtotal left hepatectomy for hepatoblastoma and was treated for ten days with parenteral nutrition and then advanced to an oral regimen. The last youngster, M.A., a 16 year old female, with post necrotic cirrhosis, membranous glomerulonephritis and chronic pancreatitis developed an abdominal abscess and necrosis of the distal pancreas following splenectomy for laceration of the spleen. She received total parenteral nutrition for 35 days before oral feedings were tolerated.

While specific indications for the use of parenteral alimentation in the management of liver disease remain unestablished, much more is known about the effect of this therapy on the normal liver. Our previous description (Cohen, Litt, Schonberg and Spigland, 1972) of abnormalities in serum transaminases, alkaline phosphatase, and occasionally bilirubin, coupled with the clinical findings of nausea, vomiting, right upper quadrant pain and hepatomegaly during a course of parenteral nutrition has now been corroborated by others (Heird, Driscoll, Schullinger, Grebin and Winters, 1972). The explanation for these aberrations may represent a direct toxic effect of these intravenous solutions on the liver. An alternative hypothesis is that these carbohydrate-protein mixtures place an abnormal burden upon normal enzymatic liver systems. Elevations in serum enzymes would thus represent an adaptive process in the metabolic conversion of these nutrients. This concept is partially supported by the repeated observations that both clinical and chemical findings revert toward normal if parenteral nutrition is continued. We would submit that these effects are mild, and in our limited experience, have thus far not been associated with life threatening disorders. Reports of other investigators suggest similar mild abnormalities (Heird et al., 1972), while one incident of an infant death occurring during intravenous alimentation with severe hepatic damage has been described (Peden, Witzleben and Skelton, 1971).

Previously reported (Cohen et al., 1972) experimental in vitro guinea pig liver explant studies from this laboratory tend to support the suggestion that the parent hydrolysate compounds and selective amino acids may be responsible for the abnormal liver function studies in children being treated with parenteral nutrition. In these studies four amino acids (glycine, leucine, isoleucine, threonine) were shown to increase serum GPT activity in organ culture but the mechanism for this elevated activity could not be further defined.

Table I describes our clinical experience with regard to liver dysfunction in 18 pediatric patients on parenteral nutrition. They ranged in age from 3 weeks to 21 years. There were 10 males and 8 females with disorders primarily representing diseases of the gastrointestinal tract. The nitrogen source in the infusion mixtures was either a fibrin or casein hydrolysate with 20 gms of glucose per 100 ml, a balanced electrolyte solution with added calcium, magnesium, phosphate and vitamins. Nausea and vomiting were encountered frequently as was right upper quadrant discomfort and hepatomegaly. Since all of these findings are not specific for liver disease and could be etiologically related to a variety of other processes, their presence cannot be solely attributed to the intravenous nutrition solutions. Nonetheless, the appearance of these signs and symptoms after the initiation of intravenous nutrition coupled with the abnormal elevation of previously normal serum GOT, serum GPT, alkaline phosphatase and bilirubin in the majority of the children, raises concern about the potential effects of parenteral nutrition fluids. Resolution of these abnormalities occurred usually within 5 days after discontinuance of the infusions. In our experience these chemical abnormalities improved if the infusate was continued although they never return completely to the normal range. Others have had a similar experience (Heird et al., 1972).

Two patients underwent percutaneous liver biopsy during the course of intravenous alimentation because of the development of the clinical and laboratory abnormalities described previously. Fig. 1 shows a liver biopsy of a 7 week old infant (J.M.) with intestinal atresia, who had been treated with parenteral nutrition for the previous 6 weeks. There is evidence of cholestasis and infiltration of the portal triad. The patient was rebiopsied at 16 weeks of age while still continuing on intravenous nutrition. A round cell infiltrate in the portal area with some fibrous tissue was noted. A third biopsy performed at six months of age while the child was still being maintained on parenteral nutrition revealed the continued presence of fibrous tissue and a round cell infiltrate in the portal triad.

TABLE 1. CLINICAL EXPERIENCE WITH INTRAVENOUS NUTRITION AND HEPATIC DYSFUNCTION

Patients	Age	Sex	Diagnosis	Duration of infusion (days)	Findings Related to the Liver		
					Clinical	Physical	Laboratory
1. A.H.	16 years	M	Granulomatous ileo-colitis	8	Nausea, vomiting	RUQ tenderness	Abnormal
2. A.I.	15 years	F	Intestinal obstruction	7/9/203	Nausea, vomiting	RUQ tenderness	Abnormal
3. J.S.	13 years	M	Abdominal abscess	12	Vomiting	None	Abnormal
4. J.M.	1 month	M	Intestinal atresia	320	Vomiting	Hepatomegaly	Abnormal
5. W.B.	14 years	M	Hepatoma	10	None	Not applicable	Not applicable
6. F.F.	1 month	M	Chronic diarrhea	8	None	Hepatomegaly	Abnormal
7. S.P.	3 months	F	Chronic diarrhea	7/19	None	None	Normal
8. M.G.	17 years	M	Ulcerative colitis	34	None	None	Normal
9. L.V.	1 year	F	Biliary atresia	14	None	Not applicable	Not applicable
10. R.I.	15 years	F	Anorexia nervosa	6	Nausea	Hepatomegaly	Abnormal
11. M.B.	18 years	M	Granulomatous ileo-colitis	22	Nausea	Hepatomegaly	Abnormal
12. M.A.	16 years	F	Cirrhosis, pancreatitis, nephritis	35	Nausea, vomiting	Not applicable	Not applicable
13. D.S.	13 years	M	Regional enteritis	42	None	None	Abnormal
14. R.G.	8 years	M	Granulomatous colitis	45	Nausea	Hepatomegaly	Abnormal
15. M.P.	21 years	F	Granulomatous ileo-colitis	11	Nausea	None	Abnormal
16. L.G.	18 years	F	Granulomatous ileo-colitis	48	None	None	Abnormal
17. I.W.	18 years	F	Granulomatous colitis	42	None	None	Abnormal
18. D.D.	3 weeks	M	Hirschsprung's Disease, enterocolitis	21	Vomiting	None	Abnormal

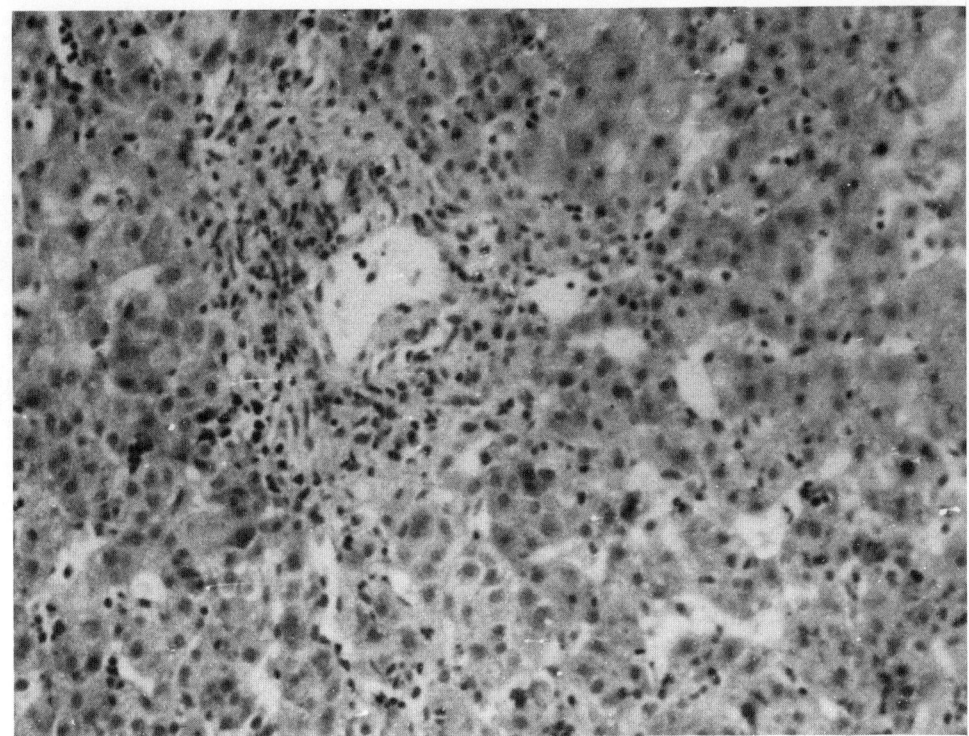

Fig. 1. Section of a percutaneous liver biopsy from patient J.M. at 7 weeks of age after 6 weeks of parenteral nutrition. Note the small lymphocytic infiltrate extending beyond the portal triad. H & E stain x 100.

L.G., an 18 year old with granulomatous ileocolitis treated for 14 days with parenteral nutrition had a percutaneous menghini needle liver biopsy after clinical and chemical abnormalities suggested early hepatic disease. Fig. 2 represents a high power view of the biopsy specimen showing evidence of a focal round cell infiltrate with mild to moderate steatosis.

Since one could not exclude the possibility that the findings in this teenager were not related to the mild focal abnormalities often seen in granulomatous ileocolitis, two subsequent patients (M.P. and I.W.) with this latter disease and in whom this treatment modality appeared indicated were subjected to a percutaneous liver biopsy prior to the initiation of parenteral nutrition. In both instances abnormal hepatic parenchyma was appreciated. Fig. 3a demonstrates the focal inflammatory cell response in M.P. Repeat biopsies of both patients after several weeks of parenteral nutrition revealed a more severe degree of these same findings. (Fig. 3b).

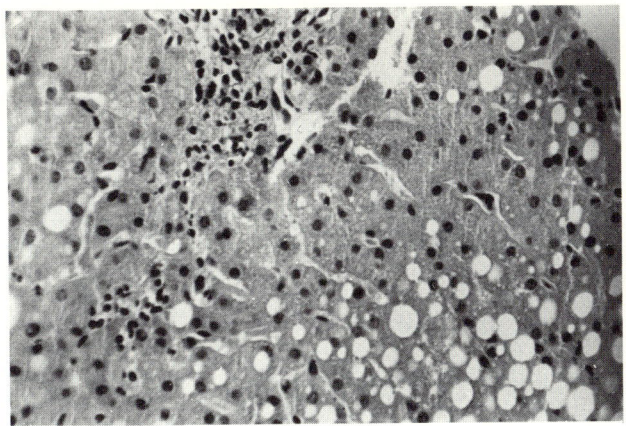

Fig. 2. Section of a percutaneous liver biopsy from patient L.G. after 2 weeks of parenteral alimentation showing evidence of focal round cell infiltration with mild steatosis. H & E stain x 120.

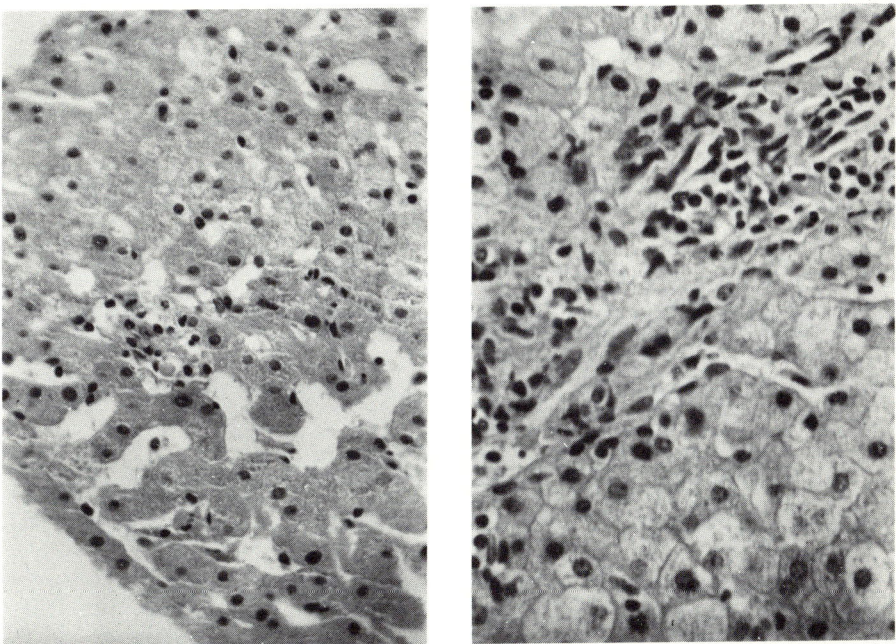

Fig. 3 a and b. Section of a percutaneous liver biopsy from patient M.P. before and after a brief course of parenteral nutrition. Very mild infiltration of hepatic parenchyma with lymphocytes prior to infiltration of therapy was most severe on the repeat biopsy (b). H & E stain x 120 and x 360.

It is difficult to correlate the clinical and chemical findings with these histological abnormalities which, in some patients, antedate the start of therapy. The apparent progression of morphologic hepatic abnormalities during parenteral nutrition suggests the need for more detailed studies of the effect of these solutions on the liver. While the number of patients we have studied is limited, our data suggest the need for a cautious approach with continued meticulous attention to potential clinical, chemical, and histological abnormalities in order to clarify this issue. Certainly, if there is a hepatoxic effect of these solutions, albeit mild, superimposing it upon an already diseased organ, as the liver often is in granulomatous ileocolitis, may be a real hazard.

CHRONIC INFLAMMATORY BOWEL DISEASE

The use of parenteral nutrition as a primary therapeutic modality for the management of chronic inflammatory bowel disease has only recently been suggested. Dudrick first observed almost total remission of granulomatous ileocolitis as a serendipitous development in patients receiving parenteral nutrition in preparation for operation (Dudrick and Rhoads, 1971). He has subsequently utilized this technique as the primary treatment in a group of patients with this disease (Dudrick, personal communication). Clinical and roentgenographic resolution occurred and the remissions have lasted for as long as 2 years. Fischer et al. (1972) recently reported 12 patients, 7 of whom had either regional enteritis or granulomatous colitis and 5 with chronic ulcerative colitis, all of whom were treated with parenteral nutrition. Both groups of investigators reported marked clinical improvement with chronic granulomatous ileocolitis, and in each series the vast majority of patients were adults.

In the past 2 years, we have used parenteral nutrition in the treatment of 7 children and adolescents with granulomatous ileocolitis and 1 boy with chronic ulcerative colitis. No benefit was observed in the youngster with chronic ulcerative colitis. Two patients, (D.S. and R.G.) have had a complete clinical remission with good weight gain, relief of abdominal cramps, defervescence of fever, and normalization of stooling patterns after approximately 6 weeks of total parenteral nutrition. There was marked improvement on radiological examination in these children (Figs. 4 and 5). The lengths of clinical remissions have been 8 and 6 months. One of these patients had never received corticosteroids, ACTH, or Azulfadine, and the other had had a recent unsuccessful single trial of prednisone and azulfadine for approximately six weeks. At the time of total parenteral nutrition, no nutrient or water was taken by mouth and no other treatment modality was utilized. The courses

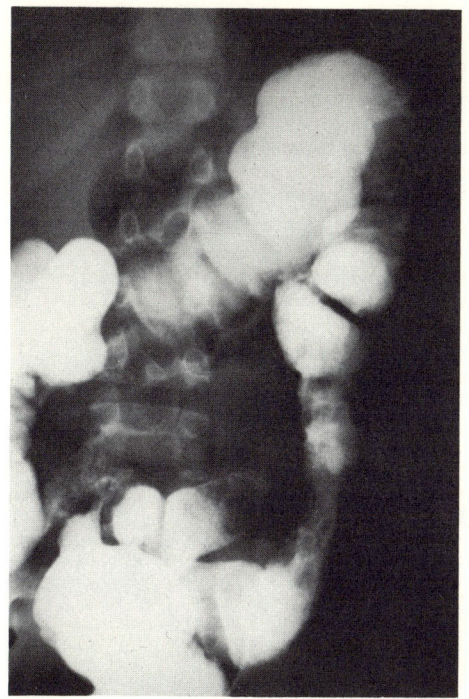 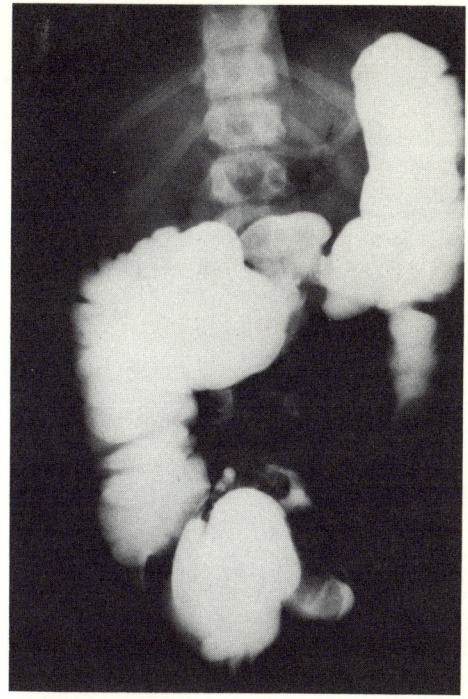

A B

Fig. 4. Barium enema of patient R.G. demonstrating the effect of parenteral nutrition on granulomatous colitis. A: granulomatous process in lower segment of descending colon prior to therapy. B: demonstrates healing 6 weeks after parenteral nutrition.

of therapy for the youngsters were 42 and 45 days.

In 2 of the patients (L.G. and I.W.) partial improvement has been noted. One 18 year old girl who had a previous small intestinal resection was admitted with small and large bowel involvement, a large inflammatory mass, diarrhea, cramps and weight loss. She received a 48 day course of parenteral nutrition with some diminution in the size of the mass, as well as incomplete resolution of her clinical complaints. Although there has been no radiologic evidence of remission, the frequency and consistency of stools have improved, her cramps have diminished and she has maintained the 10 lbs. gained during intravenous nutrition. She is currently 4 months post treatment and continues to feel well but with a persistent palpable left lower quadrant tubular mass which is slightly tender to examination. She remains free of any medication. The other patient (I.W.) showing partial improvement

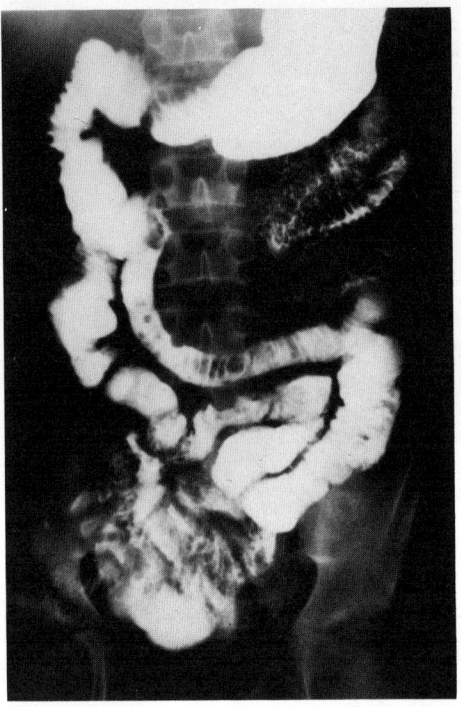 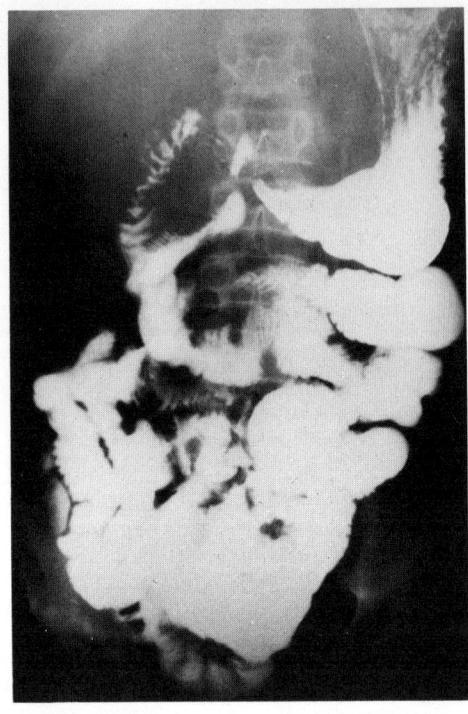

A   B

Fig. 5. Radiographs of small intestine on patient D.S. before (A) and two months after (B) a 6 week course of total intravenous nutrition. Fig. 5A demonstrates a fixed rigid loop of jejunum and Fig. 5B shows marked improvement of the upper small bowel.

presented with fever, weight loss, diarrhea and severe perianal tenderness and granulomata. While numerous complications unrelated to her primary intestinal disease made continuous intravenous nutrition impossible, during a four week period when she was free of other problems, her initial symptoms improved and a 10 lbs. weight gain was noted.

One patient (M.B.) who failed to respond to all other therapeutic modalities including prednisone, ACTH, azathiaprine, and azulfadine, had no beneficial effect from 34 days of total parenteral nutrition. In spite of surgical intervention, he succumbed from chronic and severe inanition and sepsis after a 9 month hospitalization.

In two patients (A.H. and M.P.) parenteral nutrition was employed as a supplement to surgical therapy. In one boy (A.H.)

whose primary presenting problem was intestinal obstruction and
marked weight loss, parenteral alimentation was utilized as a preoperative means of improving his nutritional state. While it
was hoped that this might also produce amelioration of the inflammatory obstruction, this did not occur. The subsequent intestinal
resection was followed by an uncomplicated post-operative period
with primary wound healing. In another girl (M.P.) with an ileal
inflammatory mass, entero-enteric fistulae and severe rectal stricture with multiple perianal fistulae, parenteral nutrition was used
following ileo-right colectomy to provide nutrition during an
11 day period of bowel rest. In this patient, an uneventful postoperative course was noted. Neither of these patients is currently
on any medication and they have remained symptom free.

Of special interest in this group of youngsters has been the
response of severe rectal and perianal disease to total parenteral
nutrition. In one boy (R.G.) who failed to respond to several
operations, his upper intestinal tract remission has been accompanied by complete healing of multiple perianal fistulae. In the
girl (M.P.) who had an ileocolectomy and post operative parenteral
nutrition, multiple perianal fistulae have healed. In this patient
the effect of resection of intra-abdominal disease must be considered, but the rapidity of the perineal improvement was similar
to that noted for R.G. Lastly, I.W. with severe anal granulomata
and ulcers sustained marked improvement of the anal involvement
after a brief course of parenteral alimentation.

This limited experience in the management of 7 children and
adolescents with clinical, radiographic, and histologic evidence
of regional enteritis or granulomatous colitis suggests that total
parenteral nutrition may have a very significant role in the primary management of this illness. While our results have not been
as successful as those reported in the treatment of adults with
granulomatous ileocolitis, they are encouraging. Our less striking responses may be related to the specific nature of the disease
during childhood and adolescence. We have the distinct impression
that the severity and aggressiveness of granulomatous enterocolitis
is greater in this age group. The value of parenteral alimentation
may merely be that of the improved nutritional state of these
debilitated patients. The weight gain and healing of fistulae may
only represent conversion to a positive nitrogen balance and enhanced healing. This, however, would not explain the reversion of
roentgenographic findings nor the protracted remissions that have
been noted. The potential ameliorating effect of total parenteral
nutrition on placing the bowel "at rest" by decreasing intestinal
secretions, and presumably reducing intestinal lymphatic flow and
congestion, remains a challenging question in establishing the
specific efficacy of parenteral nutrition in chronic inflammatory
bowel disease in childhood.

ACKNOWLEDGMENTS

Supported in part by United States Public Health Service Research Grant AM 13785-02.

REFERENCES

Boley, S.J., Cohen, M.I. and Gliedman, M.L. Pediatrics, 46, 929-933, 1970.

Cohen, M.I., Winslow, P.R. and Boley, S.J. New Eng. J. Med., 280, 1285-1286, 1969.

Cohen, M.I., Litt, I.F., Schonberg, S.K. and Spigland, I. N.I.M.C.H.D. Monograph "Intravenous Nutrition in the Newborn" 1972 (in press).

Dudrick, S.J. Personal communication.

Dudrick, S.J. and Rhoads, J.E. J. Amer. Med. Assoc., 215, 939-949, 1971.

Fischer, J.E., Foster, G.S., Abel, R., Abbott, W. and Ryan, J. Presented at Ann. Mtg. of Soc. of Surg. of the Alimentary Tract. San Francisco, June 1972 (Abstract P. 54).

Folkman, J., Philippart, A., Wah-Juntze and Crigler, J. Surgery, 72, 306-314, 1972.

Heird, W.C., Driscoll, J.M., Jr., Schullinger, J.N., Grebin, B. and Winters, R.W. J. of Pediatrics, 80, 351-372, 1972.

Peden, V.H., Witzleben, C.L. and Skelton, M.A. J. of Pediatrics, 78, 180, 1971 (Letter).

PARENTERAL NUTRITION OF RENAL DISEASE

Josef E. Fischer

Assistant Professor of Surgery, Chief of Hyperalimentation Unit, Massachusetts General Hospital, Harvard Medical School, Boston, Massachusetts

The rationale for dietary therapy in renal failure has been investigated largely in adults in chronic renal failure. Their nutritional needs have been established primarily on the basis of oral dietary intake and prolonged balance studies. The requirements for maintaining positive nitrogen balance in adults with chronic renal disease were substantiated largely through the work of Rose and Dekker (1956), Giordanno (1963), and Giovanetti and Maggiore 1964). This has led to the concept of supplying nitrogen of "high biologic activity", largely in the form of L-essential amino acids. Nitrogen from urea, which is plentiful in uremia, is then synthesized into non-essential amino acids and, with essential amino acids, into protein, the result being the maintenance of positive nitrogen balance on the 0.5 gm per kg per day of protein supplied.

In acute renal failute, a prospective study carried out at the Massachusetts General Hospital with my colleagues (Abel, Beck, Abbott, Ryan Barnett and Fischer, 1972) has revealed striking benefits following the intravenous administration of L-essential amino acids with hypertonic glucose in the therapy of acute renal failure as opposed to standard therapy consisting of hypertonic glucose alone (Abel et al., 1972). These effects include better nutrition, control of deleterious ionic excesses, (Abel, Abbott and Fischer, 1972) and most important, increased survival particularly when dialysis is necessary (Abel et al., 1972). A tantalizing suggestion, implicit in the results, and previously suggested by Dudrick (Dudrick, Steiger and Long, 1970) is that perhaps hyperalimentation in acute renal failure may even decrease the duration of acute tubular necrosis.

The previous studies carried out in this area have, by and large, not been concerned with the therapy of renal failure in children. The dietary requirements in adults, in whom growth is not essential, may be easy to satisfy, on the basis of such an L-essential amino acid mixture. Whether this applies to children in chronic renal failure is another matter. However, in children with acute renal failure, the process is often a rather self limited one, since with the exception of cardiac surgery, burns and severe trauma, pediatric age group patients with acute renal failure are rare. Our studies have not to date considered other lesions such as glomerulonephritis or nephrotic syndrome but there is every reason to believe that similar benefits might apply to those patients as well.

One potential problem in the parenteral nutrition of children with acute renal failure is the presence or absence of histidine from the mix. In the "Renal Failure Intravenous Diet" (Table 1) currently in use at the Massachusetts General Hospital (Freamine-E) histidine is not part of the mixture.

Table 1: M.G.H. "Renal Failure Intravenous Diet"

Composition/100 ml

L-Isoleucine	0.56 gm
L-Leucine	0.88
L-Lysine HCl	0.80
L-Methionine	0.88
L-Phenylalamine	0.88
L-Threonine	0.40
L-Tryptophan	0.20
L-Valine	0.65
Essential Amino Acids	5.25 gm

Each Unit Contains:

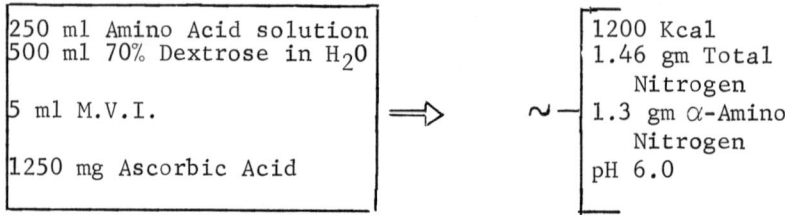

250 ml Amino Acid solution
500 ml 70% Dextrose in H_2O
5 ml M.V.I.
1250 mg Ascorbic Acid

⟹ ∼

1200 Kcal
1.46 gm Total Nitrogen
1.3 gm α-Amino Nitrogen
pH 6.0

Table 1 shows the formulation of "Renal Failure Intravenous Diet" currently in use at the Massachusetts General Hospital. This mixture contains only the 8 L-essential amino acids with hypertonic glucose in a solution containing sufficient calories so that a limited amount of fluid can be administered to patients either oliguric or anuric.

There is some doubt as to whether histidine, in fact, is required in the therapy of renal failure but at the present time the evidence does not appear overwhelming. Since the question has been previously raised as to whether histidine constituted an essential amino acid in children (Stifel and Herman, 1972), it may therefore be that the L-essential amino acid mixture in a therapy of renal failure in children should be different than that of the adult.

This, however, is merely academic since the duration of renal failure in most children is brief. The high regenerative capacity of renal tubules in the presence of normal vasculature makes a brief duration of acute tubular necrosis more likely than in adults with a similar insult. The maintenance of adequate nutrition for continued growth during an episode of acute renal failure is perhaps less important than the maintenance of some healing capacity, a decrease in the blood urea nitrogen (Dudrick et al., 1970; Wilmore and Dudrick, 1969), avoidance of dialysis, as well as control of some of the deleterious netabolites, potassium, phosphate and magnesium (Abel et al., 1972).

With these rather limited goals in mind, the following constitutes recommendations based on our experience in the therapy of acute renal failure in childhood. The beginning and optimal rates of administration of the "Renal Failure Intravenous Diet" are given in Table 2.

Table 2: Guide to Clinical Usage of "Renal Failure Intravenous Diet" (as currently used at the Massachusetts General Hospital).

CLINICAL USAGE

Rates	
Beginning	0.3-0.4 ml per hr per kg
Optimal	0.8-1.0 ml per hr per kg
Calories (optimal)	45 Kcal per kg per 24 hr
Amino Acid Nitrogen (optimal)	0.045 gm per kg per 24 hr

These constitute approximations since our experience is not extensive. The administration of approximately 0.05 gm of nitrogen per kg per 24 hrs in a child appears to be satisfactory for control of the rise in blood urea. Control of the potassium, phosphate, and magnesium ions is largely a glucose effect, aided by the presence of L-essential amino acids as previously noted (Abel et al., 1972). There is little evidence that an administration rate greater than 1 ml per kg per hr is beneficial. Indeed, in adults, more rapid flow rates may result in the further incidence of metabolic complications such as hyperglycemia (Abel et al., 1972).

The rapid ability of a pediatric kidney with its normal vasculature to recover following an insult resulting in acute tubular necrosis is manifested in the following case. In this example, the value of the "Renal Failure Intravenous Diet" helping to bring about a rapid decline in blood urea nitrogen, and thus obviate the need for a dialysis (rather difficult in pediatric patients) is seen, and confirms the value of such therapy in this age group.

Case Report

R.G.: With a known ventricular septal defect since birth, underwent pulmonary banding at the age of 10 months. At the age of 3 years, he was readmitted for total correction. Because of intraoperative and postoperative complications, the patient required pharmacologic vascular support for several days. The blood urea nitrogen rose to a maximum of 138 mg%. The "Renal Failure Intravenous Diet" was begun at a rate of 0.3 to 0.4 ml per kg per hr and gradually increased to the optimal infusion of 1 ml per kg per hr. At that time a minor improvement in renal function occurred as manifested by a drop in the serum creatinine. The blood urea nitrogen fell from 138 mg% to approximately 50 mg% in 48 hours; the "Renal Failure Intravenous Diet" was tapered rapidly.

This case provides an example of how the addition of L-essential amino acids with hypertonic glucose given by the intravenous route may obviate the need for dialysis when it appeared it might be necessary. The previously listed advantages in pediatric metabolism and the ability to recover from a massive insult at a rate much faster than the adult population makes this technique potentially valuable in the pediatric age group. Application of the techniques of intravenous hyperalimentation with the special amino acid mixes available are as valuable in this age group than they appear to be in adults. Further modifications of the available diets may be necessary to provide most efficient utilization of the amino acid needs in growing children.

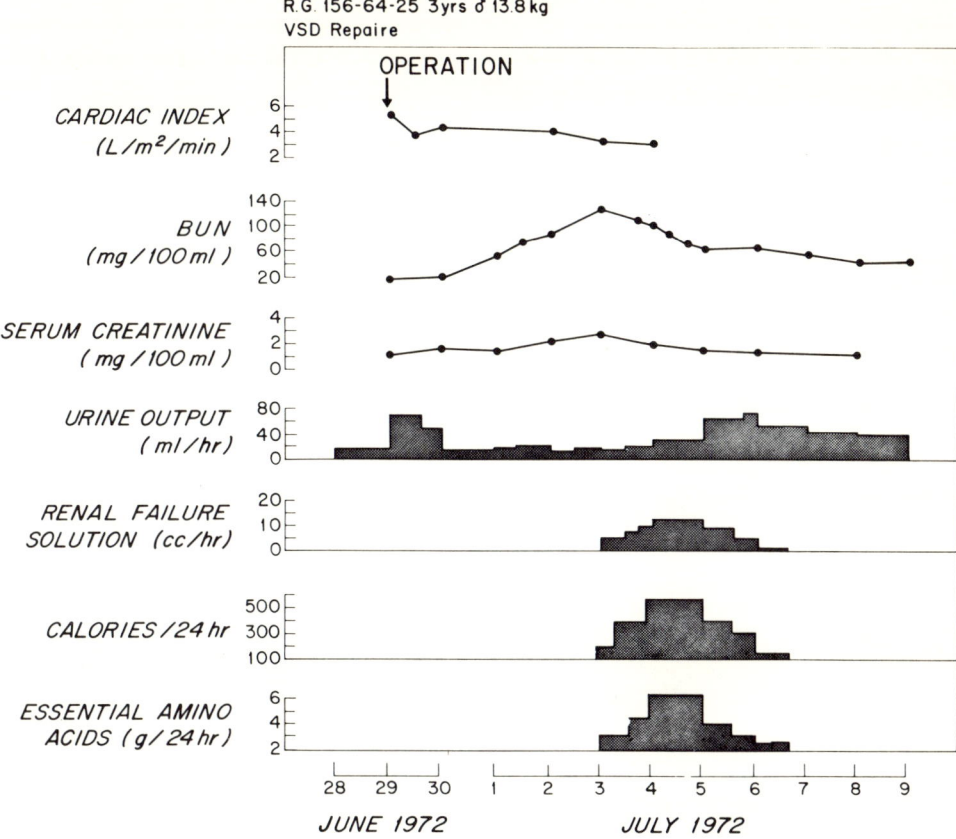

Fig. 1. Graphic illustration of the hospital course of R.G., a 3 year old patient who underwent correction of a ventricular septal defect with removal of a pulmonary band. Although renal failure appeared to be improving at the time renal failure solution was started, the rapid fall in blood urea nitrogen (90 mgm% in 48 hours) appears to indicate a beneficial effect of "Renal Failure Intravenous Diet."

SUMMARY

The application of intravenous hyperalimentation techniques using a mixture of 8 L-essential amino acids and hypertonic glucose is discussed. While experience with this mode of therapy is limited, it does appear as if this mode of therapy may be valuable. The application of the mixtures of L-essential amino acids currently in use is discussed.

ACKNOWLEDGEMENTS

We would like to thank Drs. Mortimer Buckley and Alan Goldblatt for permission to publish the case of R.G.

Freamine-E was kindly supplied by Mr. Robert Nicora, McGraw Laboratories, Glendale, California.

REFERENCES

Abel, R.M., Abbott, W.M. and Fischer, J.E. Amer. J. Surgery, 123, 632, 1972.

Abel, R.M., Beck, C.H., Jr., Abbott, W.M., Ryan, J.A., Barnett, G.O. and Fischer, J.E. Surg. Forum, 1972 (in press).

Abel, R.M., Beck, C.H., Jr., Abbott, W.M., Ryan, J.A., Barnett, G.O. and Fischer, J.E. (Submitted for publication, 1972).

Dudrick, S.J., Steiger, E. and Long, J.M. Surgery, 68, 180, 1970.

Giordanno, C. J. Lab. Clin. Med., 62, 231, 1963.

Giovannetti, S. and Maggiore, Q. Lancet I, 1000, 1964.

Rose, W.C. and Dekker, E.E. J. Biol. Chem., 223, 107, 1956.

Stifel, F.B. and Herman, R.H. Amer. J. Clin. Nutr., 25, 182, 1972.

Wilmore, D.W. and Dudrick, S.J. Arch. Surg., 99, 669, 1969.

PARENTERAL NUTRITION IN CRITICAL ILLNESS

John T. Herrin

Department of Pediatrics, Harvard Medical School

Massachusetts General Hospital, Boston, Massachusetts

In many patients with critical illness the standard approach to parenteral alimentation is not possible, for the limits of water, sugar and electrolyte tolerance are dependent on the underlying condition. Furthermore, the growth process in an unstressed infant allows incorporation of potassium, phosphate, water and nitrogen into newly formed cells so that smaller solute and water loads are presented for excretion. It is unreasonable to expect average anabolic requirement to apply in critical illness where growth is not occurring.

The priority for establishment and maintenance of an adequate circulation and urine output, followed by replacement or repair of metabolic abnormalities often requires interruption of a feeding regime or delays its commencement. This series of priorities thus restricts the delivery of a high caloric load until stabilization has occurred. During this early time we aim to provide a basal caloric supply as high as is practical to prevent metabolic deterioration. Relatively low amounts of carbohydrate are sufficient to minimize protein loss and prevent ketosis. After stabilization has occurred it is possible to progress to higher levels required for growth.

Problems

Central intravenous alimentation has been associated with many problems: septicemia (Boeckman C.R. et al., 1970), thrombosis and thrombophlebitis (Groff, 1969), hypersensitivity reaction (Rea et al., 1970), air embolism (Filler et al., 1970), and hyperosmolar coma (White et al., 1972). Any of these complications markedly increases the risk to the critically ill patient. It is for this reason mandatory that circulation, electrolyte and acid-base status be stabilized prior to the commencement of central venous nutrition. A scheme for priorities used in our patients is outlined in Table 1. Limits for tolerance are determined by monitoring serum and urinary electrolyte concentration, osmolality, glucose, urea and amino acid concentration. Our aim is to administer the highest possible caloric load while maintaining the above parameters within normal physiological range.

Table 1: Priorities in Acute Care of Critically Ill

1. Maintain Circulation

2. Establish and maintain urinary output

3. Correct existing fluid, electrolyte and acid-base disturbances

4. Administer necessary drug therapy, e.g. antibiotics

5. Caloric supplementation

For monitoring of such vital functions as circulation, ventilation and acid base status in the critically ill child, indwelling venous or arterial catheters are necessary. To maintain patency of these catheters and to allow blood sampling, a small but finite volume of fluid is necessary. For optimal safe usage, isotonic saline or 5% dextrose in water solution is administered in a low constant volume and intermittent flushing with small bolus of fluid is carried out as necessary to maintain patency of the line. The caloric value of these fluids is low compared to standard parenteral alimentation solutions, yet in the very small infant, the volume constitutes a significant proportion of absolute water need.

Table 2 outlines common conditions resulting in changes in water requirements encountered in the intensive care situation. In general terms water need is reduced and therefore fluid restriction is required in most patients with critical illness. The increased requirement in polyuric renal failure, fever, gastrointestinal losses or after extensive surgery pose a lesser problem. In practice the delivery of adequate calories may be facilitated.

Table 2:   Alteration in Water and/or Salt Administration

a) Limitation to below normal tolerance
   i) Organ failure - renal (oliguric)
                    - cardiac
                    - respiratory
                    - liver
   ii) C.N.S. disease - encephalitis
                      - meningitis
   iii) Inappropriate Antidiuretic Hormone (A.D.H.) Secretion
   iv) Therapy - positive pressure respiration
               - humidification
               - control of ambient temperature
               - decreased movement
               - drugs (morphine)

b) Increase tolerance of water and/or salt
   i) polyuric renal failure - acute and chronic
   ii) surgical patient
   iii) fistulae or gastrointestinal losses
   iv) fever

Table 3:   Primary Conditions

Respiratory Distress Syndrome	32
Prematurity	7
Postcardiac surgery	5
Head injury	3
Encephalitis-status epilepticus	2
Polyuric renal failure	3
Malabsorption syndromes	2
Bowel surgery prior to refeeding	3
Tracheo-esophageal fistula	1
Regional ileitis	2
Metabolic disease - hyperglycinemia	1
- Maple Syrup Urine Disease	1
Failure to thrive - emotional deprivation	1
- rubella syndrome	1
Malignancy - during tumour therapy	
- lymphosarcoma	1
- neuroblastoma	1
Total number	66

Clinical Experience

Our experience in 66 patients is described in Table 3. Ages ranged from 4 days to 15 years. Calories were administered intravenously using solutions outlined in Table 4. Total intravenous nutrition was administered centrally in 16 patients and by peripheral infusion in 3 patients. Thirty-seven patients received supplemental calories by peripheral infusion. In the supplemental technique used in our unit, concentrated oral formula is gradually increased to gastrointestinal tolerance using small volume feedings administered by nasogastric tube. Peripheral intravenous administration of 5% dextrose, 10% dextrose or "Pedihypercal" (15% glucose-fructose, see Table 4) were used to supplement caloric intake, provide necessary water and electrolyte requirements and maintain patency of monitoring lines. Such a supplemental technique obviates some of the problems of restriction of ability to provide adequate calories within the available volume for maintenance in critical illness, particularly in very small infants.

Average calories delivered varied between 35 and 160 Kcal/kg/day (average 80 Kcal/kg/day). The large variation was occasioned by limitation of water and electrolyte in different disease states. Volumes of fluid required to provide these caloric loads ranged between 35-90 ml/kg/day.

Illustrative Cases

The following series of short cases illustrates the choice of route, potential gains and problems as they occur in patients with critical illness.

CASE 1

An eight month old child was admitted for inadequate weight gain (only two pounds over birth weight). Oral feeding of standard formula at 150 cal/kg in hospital was apparently well tolerated but little weight gain followed. A peripheral line was placed and using "Pedi-hypercal," Aminofusin (Pfrimmer) and 10% Intralipid (Vitrum). Caloric intake was advanced to 120-130 cal/kg with weight gain.

Comment: In this child total caloric requirement was supplied by peripheral hyperalimentation, thus avoiding the risks and hyperosmolar limitations of central administration. No indwelling monitoring lines were required. Normal water tolerance facilitated administration.

Table 4

Preferred Route of Administration	Indication	Feeding Type	Calories/ml
A. Peripheral	sepsis - burns	Aminofusin L600 (Pfrimmer)	
	potential skin sepsis	Pediatric Hypercal[1]	0.53 cal/ml
	epidermolysis	Pediatric Hypercal with alcohol[2]	0.64 cal/ml
	exfoliative dermatitis	Intralipid 10% (Vitrum)	1.1 cal/ml
B. Supplemental	prematurity	Pediatric Hypercal	0.53 cal/ml
	respiratory distress	Dextrose 10%	
	organ failure	Concentrated oral formula - Ross Similac PM 60/40 low Na formula	1.0-1.16 cal/ml
C. Total Parenteral	intact skin site		
	surgical patient bowel obstruction fistulae chronic suppurative disease	M.G.H. hypercal solution[3]	0.93 cal/ml
	polyuric renal failure		
	malabsorption states		

1. Dextrose 8.5, fructose 6.5, pH 7
2. Dextrose 8.5, fructose 6.5, pH 7, 2% ethanol
3. See page 226

CASE 2

A one day old, 1.4 kg premature infant presented with increasing respiratory distress. Therapy required included high inspired oxygen concentration (FiO_2) and positive end expiratory pressure (P.E.E.P.) respirator therapy. By day three control of oxygenation on FiO_2 0.40 and 4-6 cm H_2O P.E.E.P. was associated with inappropriate antidiuretic hormone secretion. Serum osmolality fell to 275 mOsm/kg despite limitation of fluid administered to 50 ml/kg water/24 hours. This was administered as "Pedihypercal" through an umbilical artery catheter. Limitation of administered water to 30 ml/kg/24 hours was followed by an increase in serum osmolality to 290 mOsm/kg and improvement in gas exchange.

CASE 3

A one day old premature 1.2 kg infant was admitted with respiratory distress. Control was obtained using high FiO_2 and P.E.E.P. respiration. Gases were monitored using an indwelling right radial artery line. When it became obvious on day five that long-term respirator therapy would be required, a central line was placed for hyperalimentation. The combined water load was then 40-50 ml/kg/day with "central feeding" limited to 20-25 ml/kg/day. With this water load serum osmolality fell to 270 mOsm/kg with a coincident deterioration in gas exchange. Diuretic therapy and restriction of fluid was followed by restoration of the prior ventilatory status and good oxygenation, but caloric intake was reduced to 15-20 cal/kg since a volume of fluid was mandatory to maintain the radial artery line for monitoring.

CASE 4

A two day old 1.3 kg premature infant with respiratory distress was controlled using P.E.E.P. alone. After the initial 24 hours of control, small volumes of concentrated feedings (2-4 ml) were administered by gavage and the caloric value gradually increased to 0.9 - 1.1 cal/ml. "Pedihypercal" was administered through the umbilical artery catheter at 10 ml/kg/day to provide both calories and additional water to allow for the excretion of the additional solute load presented by the concentrated feedings.

Comment: In these premature infants with idiopathic respiratory distress syndrome (RDS) many problems are present. The decrease in insensible water loss resulting from humidification compounds the problem of inappropriate vasopressin secretion associated with positive pressure respirator therapy (Kjekshus et al., 1972). Fluid and caloric supply is limited by these factors.

It can be appreciated from Case 3 that the need for a monitoring line limited the ability to administer calories. When the sugar concentration of the solution was increased in an attempt to provide calories, both a hyperosmolar state and an osmotic diuresis were problems. In Case 2 the use of "Pedihypercal" solution (as outlined in Table 4) through the arterial line was well tolerated and allowed administration of approximately equal caloric loads.

We have used "Intralipid" (Vitrum) as a caloric source in three babies with RDS but discontinued its use following thrombocytopenia. It may be that the mild hypoxemia resulting from the underlying disease process led to difficulties in utilization as had previously been reported or that intercurrent sepsis present in two children and unproven, but suspected, in the other was the cause.

Case 4 shows that the milder case of RDS with easy control on positive end expiratory pressure therapy may be given adequate calories over a long-term period using concentrated formulae orally and peripheral administration of any extra water requirement either through a standard intravenous infusion or through the monitoring line using isotonic solutions.

If the gut tolerance is for only very small volumes, we have supplemented 0.9 cal/ml - 1.0 cal/cc feeding with the addition of medium chain triglyceride oil 0.5 - 1.0 ml per feeding.

CASE 5

Metabolic disease. Maple Syrup Urine Disease. This eighteen month old child was in good control until a metabolic crisis followed an intercurrent infection. He rapidly developed severe acidosis, circulatory impairment and mild dehydration. Therapy was commenced by correcting the acidosis and then simultaneously administrating an intravenous amino acid solution low in leucine, isoleucine and valine, and maintaining control of the acidosis.

Comment: In this crisis situation two major problems were present: (i) severe acidosis, the result of inability to utilize the branched chain amino acids leucine, isoleucine and valine resulted in organic ketoacidosis; (ii) suppression of ability to metabolize sugar and amino acids as a result of severe acidosis. The process is thus cyclic and therapy requires both restriction of branched chain amino acids and correction of the acidosis.

In this child this was accomplished by using a solution of specially prepared amino acids. When oral supplementation is possible, a gelatin based protein diet, 0.5 - 1.0 gm per kg/day, will supply this requirement and then other caloric sources can be administered intravenously as required. We have used total hyperalimentation in a number of infants with neonatal bowel obstruction using an approach essentially similar to that outlined by others.

Summary and Conclusions

In critical illness suitable adjustment to administered fluid load must be made to allow for changes in insensible loss and mode of therapy.

Our aim is to maintain normal metabolic conditions at all times throughout the course of an acute illness and provide the maximum calories possible within the limits imposed by salt and water tolerance.

A single indwelling arterial or venous line may be used for feeding and monitoring purposes provided that supplemental feeding techniques are used. If total intravenous nutrition is undertaken, the sterility precautions require that a second separate access line be placed for monitoring purposes.

A guideline to choice of route of administration is outlined in Table 4. The central line route for total parenteral nutrition should be limited to patients with intact skin at the site of insertion and uninfected skin surfaces. This technique requires that little or no limitation to water administration be present. Peripheral techniques using amino acid solutions, fat or supplemental feeding technique are generally safer and as efficient for delivery of calories as central administration in the intensive care of critically ill patients. Acceptance of low caloric intake is often necessary in the early stages of critical illness but basal caloric needs should be administered to minimize catabolism even in early stages of the illness.

REFERENCES

1. Boeckman, C.R., Krill, C.E Jr. J. Pediatr. Surg. 5:117, 1970.

2. Filler, R.M., Eraklis, A.J. Pediatrics 46:456, 1970.

3. Groff, D.B. J. Pediatr. Surg. 4:460, 1969.

4. Kjekshus, J.K., Mjøs, O.D. J. Clin. Invest. 51:1767, 1972.

5. Rea, W.J., Wyrick, W.J., McClelland, R.N. et al. Arch. Surg. 100:393, 1970.

6. White, W.A., Bergland, R.M. J. Neurosurg. 36:608, 1972.

PARENTERAL NUTRITION IN CHILDREN WITH BURNS

The prolonged severe stress and chronic hypermetabolism present in all patients with severe thermal injuries makes provision of adequate nutrition of paramount importance. The calories required to maintain nutritional balance and to allow wound healing in the burn patient greatly exceed basal metabolic requirements. A general guideline for the estimate of calorie requirements of burned children is 60 calories per kg per day plus an additional 30 calories for each 100 cm^2 of burned area. Burned children are often anorexic, have delayed gastric emptying, and develop diarrhea with duodenal tube feeding. Severe malnutrition is therefore a frequent complication (Blocker et al., 1955; Levenson et al., 1945; Sutherland and Batchelor, 1968) of patients with burn injury and is associated with protein breakdown, delayed wound healing and infection which further increase caloric demands.

The advent of parenteral nutrition has opened new avenues for therapy and prevention of malnutrition in these children. However, the technique of parenteral nutrition in burns poses several special problems. Because of the nature of the burned wound, sepsis is a particular hazard. Placement of the catheter is difficult especially when the burns involve the upper trunk and the arms. Whirlpool debridement may compromise sterility at the site of the catheter entrance. On the other hand, infusion of calories through peripheral veins limits the caloric concentration of the infusate and require large volume to meet nutritional needs. There is also an increased incidence of venous thrombosis whenever the dextrose concentration of the infusate exceeds 10%. Some of these complications can partially be prevented by simultaneous infusion of fat emulsion which in addition to their high caloric content also appear to protect peripheral veins from thrombosis. The reported experience with parenteral nutrition in burned children is limited and critical reviews are nonexistent. We have therefore presented the experience with parenteral nutrition of both the Cincinnati and Boston Units of the Shriners Burns Institute.

EXPERIENCE IN THE SHRINERS BURNS INSTITUTE IN CINCINNATI

Martin P. Popp, Edward J. Law, and Bruce G. MacMillan

Department of Surgery, University of Cincinnati Medical Center

Method

Twenty-six patients with either 40% full-thickness burns, or those with lesser burns and a progressive weight loss, were selected for intravenous nutritional supplementation. Their burns were treated by the exposure method, with topical antibiotics consisting of Sulfamylon, silver sulfadiazine or gentamicin. Eschar was debrided as it separated, and wounds were covered with porcine heterograft Early autografting and closure of the burn wound was accomplished as soon as the wound was ready.

Table 1. Nutrient Solutions

	Amino Acid Solution (Freamine) 500 cc.	Protein Hydrolosate 5% (Aminosol) in D5/W 750 cc.
Dextrose 50%	500 cc.	350 cc.
NaCl	50 mEq.	40 mEq.
KCl	40 mEq.	30 mEq.
Ascorbic Acid	100 mg.	100 mg.
B Complex	2.0 cc.	2.0 cc.
	1.18 Kcal/cc.	0.82 Kcal/cc.
	6.25 g. N/1000 cc.	4 76 g. N/1000 cc.

Nutritional supplementation was begun as soon as patients recovered from acute fluid imbalances, or whenever weight loss became apparent. Children were encouraged to eat as much of a high calorie and high protein diet as possible while supplementation continued. The solutions used consisted of either 5% protein hydrolisate (Aminosol) or 8.5% crystalline amino acid solution with 50% dextrose (Freamine) electrolytes, and vitamins (Table 1). They were prepared on the ward by the nursing staff and administered immediately. Rate of administration was pushed to the highest levels possible without excessive urinary output or congestive heart failure.

The solutions were administered through standard microdrop intravenous infusion apparatus without pump or filters in all but three patients. In these patients, a Harvard peristaltic pump and 22 micron filtration system was used. Central venous catheters were placed percutaneously in the subclavian vein through unburned skin in 15 patients. The other 11 patients had extensive burns over the subclavian areas. Four patients were treated with subclavian catheter through burn wound with Sulfamylon dressing, and cut-down catheters through peripheral unburned skin in jugular, basilic or saphenous veins were used in the other seven cases. The standard infra-clavicular subclavian puncture technique, as described by Dudrick, was used (2). Chest X-rays followed all catheter placements. The children were weighed every three days, as long as movement was not limited by fresh skin graft. Serum electrolytes, FBS, BUN, Ca, PO_4, magnesium, proteins and osmolality, were determined every three days and whenever indicated. These values were used to modify electrolyte composition of nutritional solutions. Urinary sugars were monitored with test tape every six hours and insulin given when glucosuria existed.

Cultures of blood, burn wound, and urine were routinely taken every three days. When signs of sepsis existed, standard cultures were taken, nutritional fluid was cultured, and the entire intravenous administration apparatus was replaced down to the catheter. If signs of sepsis persisted, the catheter was removed and cultured.

Results
26 children were treated with burns which averaged 37% full-thickness and 52% total surface. They ranged in age from 18 months to 14 years. 15 of these children gained or maintained weight while being treated, eight were limited to less than 6% weight loss, while three had appreciable weight loss of greater than 15%. The average duration of therapy was 24.6 days, with the longest periods being 68 and 59 days. The average quantity of therapy was 49.6 cc per kg per day. This would be the equivalent of 3500 K cal. and 21 gm. N. in a 70 kg adult.

There were six children who could not eat during their therapy. This included two patients who had superior mesenteric artery syndromes demonstrated by cine-radiography, and four children with duodenal ulcers demonstrated on upper gastrointestinal X-rays. Five of these children survived, and their parenteral nutrition was considered life-saving. The sixth patient required a surgical repair of her duodenal ulcer and died in terminal sepsis after a weight loss of 15.7%.

Pathology

There were seven patients who died, all with large burns which averaged 58% full-thickness. The largest burn to survive at the time of this series was a 65% full-thickness burn. Postmortem examinations were performed on all patients, and five patients were found to have septic thrombi at the point of central venous catheterization. Four of these were from subclavian or jugular catheters and propagated down the superior vena cava (Plate I). The fifth thrombus originated from a proximal saphenous vein cut-down and caused complete thrombosis of the inferior vena cava and iliac veins (Plate II). Microscopically, these were organized fibrin thrombi which were colonized with bacteria and yeast (Plate III).

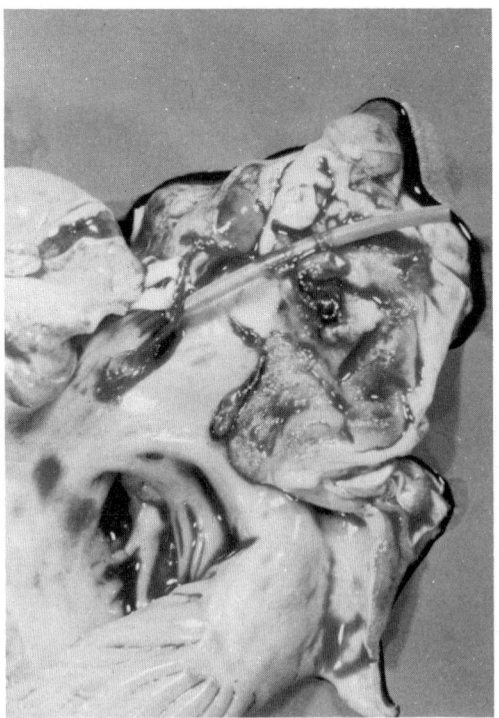

Plate I. Septic thrombus totally occluding superior vena cava with catheter still in place.

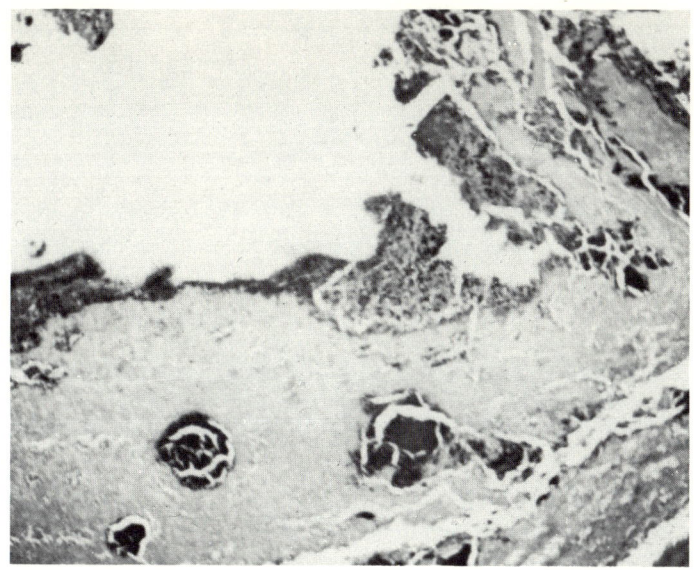

Plate II. Organized fibrin thrombus showing bacteria. Grocott stain, 400 X.

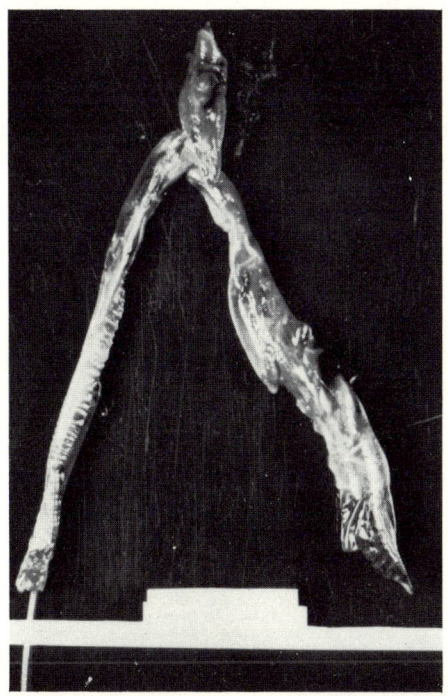

Plate III. Inferior vena cava thrombus.

Table 2. Relation of Pathology to Mode of Central Venous Catheterization

Number Patients	Type Catheterization	Deaths	Septic Thrombus
15	Subclavian through unburned skin	1	0
4	Subclavian through burn wound	3	2
	Cutdown catheters:		
5	a. Jugular vein	2	2
1	b. Saphenous vein	1	1
1	c. Basilic vein	0	0

The relationship of the thrombus formation to type of catheterization is illustrated in Table 2. In 15 patients with subclavian catheters through unburned skin, there were no known episodes of thrombosis. Two of four patients with subclavian catheters through burnwound developed thrombi, while three of seven patients with cut-down catheters through unburned skin had postmortem evidence of thrombus.

Other postmortem findings included tricuspid valvular lesions in all four patients who had been treated with superior vena cava catheters and developed septic thrombi. These lesions consisted of tricuspid endocarditis in three instances and one case with a septic tricuspid infarct. All patients had radiologic evidence of catheter in the heart at some time during their nutritional therapy. There were three patients with multiple small pulmonary emboli, but no instance of massive pulmonary embolus was noted.

Sepsis

There were 19 distinct episodes of sepsis during supplemental parenteral nutrition, each with at least three positive blood cultures for the same organisms. The most common organisms cultured were Staph. aureus and Pseudomonas aeruginosa, followed by Candida species. There were eight total episodes of the 19 in which Candida albicans or Candida species were cultured. Four of the seven deaths resulted from generalized Candidiasis.

14 of the 19 total episodes of sepsis appeared related to the central venous catheterization (Table 3). In nine of these fourteen episodes, removal of the catheter resulted in prompt clinical improvement and sterile blood cultures. In the other five of these fourteen episodes, sepsis was terminal, and septic thrombus was found at postmortem examination. Five of the nineteen total episodes were not catheter related. Three of these

were not affected by catheter removal, and blood cultures became sterile with wound care and systemic antibiotics. The other two episodes were terminal and not influenced by removal of intravenous catheters, and no sign of phlebitis was found at postmortem examination.

Table 3. Probable Etiology of Sepsis

A. Catheter Related - 14 patients
    9 responded to catheter removal
    5 terminal with postmortem thrombus
B. Catheter Unrelated - 5 patients
    3 not affected by catheter removal
    2 terminal without postmortem thrombus

Discussion

The nutritional results, with only three of 26 patients experiencing expected weight loss, demonstrate the benefit of intravenous nutritional supplementation in reducing weight loss. In addition, most clinicians felt that the patient's general clinical condition was improved during periods of supplementation. Unfortunately, nitrogen and other types of balance studies were impossible because of the difficulty in measuring losses in burn wound exudate. The quantity of supplementation given (the equivalent of 3500 K cal and 21 gm. N. in a 70 Kg. adult) does approach the levels estimated by Soroff, Pearson and Artz, to be necessary for equilibrium in adults with slightly smaller burns (3). The recent work of Wilmore, Curreri and Spitzer, with adult burns, in which the respiratory quotient was shown to shift from 0.8 to 1.01 during periods of intravenous nutritional supplementation with carbohydrate solutions is excellent evidence that the intravenous nutrients are being used to spare normal tissue.

The five episodes of septic phlebitis at central venous catheterization sites is the deterrent to successful central venous supplementation. This complication occurred only in the group of patients burned over the subclavian areas, which also included those patients with the largest burns. The results gained with placement of the subclavian catheter through burn wound, and with placement of cut-down catheters through clean skin were equally disappointing. The finding of rapid regression of sepsis when catheters are removed, however, strongly suggests abandoning central therapy when signs of sepsis present, or rotating catheters. The newer techniques during sepsis in the future. The higher levels of supplementation are possible with central intravenous techniques, however, will allow the central techniques to retain some value.

The association of septic heart lesions in patients with septic thrombi and catheters in the heart suggests either direct mechanical valvular damage by the catheters or hyperasmolar damage. This complication should be easily preventable if catheters are radiographically placed in the superior vena cava. This will also prevent perforation of the heart, as described by Fitts, et al.(4).

The problems of glucosuria and hypernatremia were encountered periodically in this series. Hypernatremia was easily corrected by slowing IV's and giving free water, while glucosuria responded promptly to giving insulin and slowing IV's.

The incidence of Candidiasis reported in this series is disturbing because of the recent association of Candidiasis with parenteral nutrition, by Ashcraft and Leape (5). The work of MacMillan, Law and Holder, however, documents the increasing problem of Candida in the burned child (6). Because of the number of variables involved, it is difficult in this series to assess the relationship between Candidiasis and central venous nutritional supplementation. The clinician caring for this type of patient, however, should be aware of this etiologic possibility when sepsis presents and treat when present by removing catheters and giving Amphotericin B.

The possibility of lethal septic thrombophlebitis indicates that central venous nutritional supplementation be used only in large burns in which nutritional deficiencies are expected, and those smaller burns which demonstrate marked weight loss while receiving maximal intestinal alimentation. With proper care and recognition of potential problems, central venous nutritional supplementation can be a valuable adjunct to burn therapy.

Summary

1. Study of 26 patients with severe thermal burns has demonstrated that intravenous nutritional supplementation can reduce expected weight loss.
2. No serious problems were encountered in 15 patients with subclavian intravenous catheters through unburned skin.
3. Septic thrombophlebitis has been demonstrated in 5 patients with very large burns including subclavian areas.
4. Rotation or prompt removal of catheters during periods of sepsis is indicated for prevention of septic thrombophlebitis.
5. The preferred method of administration in patients with burns over subclavian areas is yet to be determined.
6. Intravenous nutritional supplementation should be considered for all burns of greater than 50% full thickness and smaller burns with severe weight loss.

EXPERIENCE IN THE SHRINERS BURN INSTITUTE IN BOSTON

Alia Antoon and Hans Henning Bode

Department of Pediatrics, Harvard Medical School
Boston, Massachusetts

The results of supplementary parenteral nutrition in twelve children between the ages of two to fourteen years with severe thermal injuries are reviewed. Ten of these patients suffered third degree burns involving 50 to 90% of the total body surface. One of the children with a 30% burn had delayed healing due to severe malnutrition prior to his transfer to this hospital, i.e., loss of 50% of body weight. Another, four year old patient with 30% burns suffered anoxic brain damage and was unconscious for a period of six weeks.

In nine patients a central venous catheter had been placed for indications other than intravenous feeding. These central venous lines were then used for parenteral alimentation. As soon as the catheters were no longer considered essential for their care, supplementary nutrition was given through a peripheral vein in two of these children. In three additional patients the latter route was used entirely. Infusate for the central venous catheters was a hypertonic glucose solution containing 5% protein hydrolysate and 900 calories per liter ("MGH 900", see chapter by Dr. Fischer). For the peripheral infusion a 10% fat emulsion (Intralipid^R, Vitrum) was used in combination with a solution containing 10% sorbital and 5% L-amino acids (Aminofusin 600^R, Pfrimmer) and a 15% glucose-fructose solution, pH 7.0, prepared by our hospital pharmacy.

RESULTS

Due to the extent of the burns in some of these patients, the central venous catheters were inserted through the burned areas. Recurrent episodes of sepsis necessitated frequent changing of catheter sites. Out of the seven patients who received only central hyperalimentation, six died. Catheter sepsis contributed to the death in four of these cases. Post-mortem findings showed multiple septic emboli and mural thrombi at the catheter tips. One of the patients ruptured the vein at the catheter site due to local infection.

A two year old girl with 90% third degree burns survived. She had received central hyperalimentation for a period of two months. When catheter sepsis continued to be a major problem, she was treated by peripheral alimentation using 10% Intralipid^R, Aminofusin 600^R and the hypertonic carbohydrate solution for a period of four weeks. This allowed intravenous provision of calories to 2,500 calories per square meter body surface per day. Her infection was controlled with the change from central to peripheral venous alimentation. Three other patients who were treated only with peripheral alimentation showed no evidence of infection that could be related to the intravenous therapy.

CONCLUSION

The results presented emphasize the dilemma that faces the physician caring for patients with extensive burns. The initial treatment of shock and the constant need for plasma and blood replacement demand a maintenance of an intravenous line. There remains, however, serious doubt if the benefits of central venous nutrition ever exceed the potential complications. Even in the patient in whom the catheter insertion sites have been spared from thermal destruction, there is frequent bacteremia during debridement and other procedures leading to the bacterial invasion of the thrombus at the catheter tip.

Our limited experience with the peripheral venous nutrition on the other hand is encouraging. The simultaneous infusion of fat emulsion and hypertonic solution containing carbohydrates and L-amino acids made it possible to provide almost the entire caloric needs intravenously. The infusion sites remained patent for as long as ten days and no major problem such as thrombophlebitis or sepsis was seen. The availability of fat emulsions has made such peripheral venous nutrition possible. It remains still to be seen if the pigment changes in the RES caused by Intralipid^R interfer with the normal physiological function of this system.

Bibliography

1. Ashcraft, K., and Leape, L. JAMA, 212:454.

2. Blocker, T.G., Levin, S.L., Nowinski, W.W., Lewis, S.R., and Blocker, V. Ann. Surg.,141:589, 1955.

3. Dudrick, S.J., and Wilmore, D.W. Arch. Surg.,98:256, 1969.

4. Dudrick, S., Wilmore, D., Vars, H., and Rhoads, J.E. Ann. Surg., 169:974, 1969.

5. Fitts, C., Barnett, L., and Webb, C. J. Trauma, 10:764, 1970.

6. Levenson, S.M., Davidson, C.S., Lund, C.C., and Tayler, F.H. Surgery, Gyn. and Obs.,80:449, 1945.

7. MacMillan, B., Law, E., and Holder, I. Arch. Surg., 104:509, 1972.

8. Soroff, H., Pearson, E., and Artz, C. S.G. & O., 112:159, 1961.

9. Sutherland, A.B., and Batchelor, A.A. Ann. N.Y. Acad. Sci., 150:700, 1968.

10. Travis, S.F., Sugerman, H.J., Rubert, R.L., and Dudrich, S.J. N. Eng. J. of Med., 285:763, 1971.

11. Wilmore, D., Curreri, W., Spitzer, K., and Pruitt, B. S.G. & O., 132:881, 1971.

EXPERIENCE WITH CENTRAL VENOUS CATHETERS

C. Burri and H.H. Pässler

Department of Traumatology (C. Burri)

University of Ulm (Germany)

The indications for the use of vena cava catheters are measurement of central venous pressure, infusions and transfusions in critical circulatory conditions, long term infusions, parenteral nutrition and administration of hypertonic solutions. While the use of caval catheters may save many lives, it is not without risks. In a one year prospective study performed in nine hospitals in Germany, Austria and Switzerland* we evaluated our experience with 3241 central venous catheterizations and compared the results with other reported series (11,000 cases).

Patient material

The age distribution of our patients ranged from infancy to age 100 with most cases falling between 55 and 75 years of age. The majority of our patients (77%) were in poor health; 20% suffered from heart disease; 8% had diabetes mellitus; 3.3% were uremic; 17% had a tracheostomy or were intubated for longer than 24 hours; 13% were in shock and 3% showed signs of sepsis at the time the catheter was introduced.

Method of catheter placement

The catheters used consisted of polyvinyl chloride and were introduced percutaneously in 74% and by venous cut down in 26% of patients. Three different methods of catheter care were used:
1) Sterile dressings without any spray-application
2) Sterile dressings with Nobecutan-Spray
3) Sterile dressings with Polymycin-bacitracin-Spray

---

* Universities of Basel, Berlin, Erlangen, Mainz, Ulm, Vienna and Hospitals in Baden, Linz and Nuremberg

The three approaches to the superior vena cava recommended were
1) The external jugular vein
2) The basilic vein
3) The subclavian vein

In our opinion, the introduction of a central venous catheter through a femoral vein should be avoided, since this approach carries too many dangerous complications.

Results

In 94 patients it proved impossible to place the catheter through one of the above-mentioned sites, in most of these cases the cephalica vein was used. Catheters were left in place from several hours to 156 days with most used from 4 to 14 days. In 6% of our cases the catheter was removed prematurely by the patient and in 2% erroneously by the staff. In 12% the catheter had to be removed prematurely because of complications.

We consider the only correct position of the catheter tip to be the vena cava superior adjacent to the entrance to the right atrium and not the right atrium itself or even the right ventricle.

The three different sites of entry were evaluated for ease of final catheter placement. The first puncture of the external jugular vein failed in 32%, of the subclavian vein in 28%, and of the basilic vein in 15% of the patients. The introduction of a catheter after the successful puncture of the vessel was easiest via the subclavian vein. Jugular and basilic veins often presented difficulties in advancing the catheter, resulting in poor positioning of its tip.

Serious complications during placement of the catheter were limited to the subclavian vein. The incidence of injury to the subclavian artery in our series was 1%, which corresponds to the 1.6% reported in the literature. Pleural injury was observed in 0.8% of our cases while this complication occurred in 1.2% of the previously reported series. Training and careful technique did reduce the frequency of this complication.

We found nine reports of perforated central veins and nine cases of the cardiac perforation in the literature. Plexus lesions occurred in three cases and air embolism in eleven patients. These complications caused a high rate of mortality. Equally dangerous are catheter embolism and these are more frequent than can be assumed from literature. Based on previously reported cases, a questionnaire and the presently reported prospective study, we were able to collect 214 incidents of catheter embolism (Fig. 1).

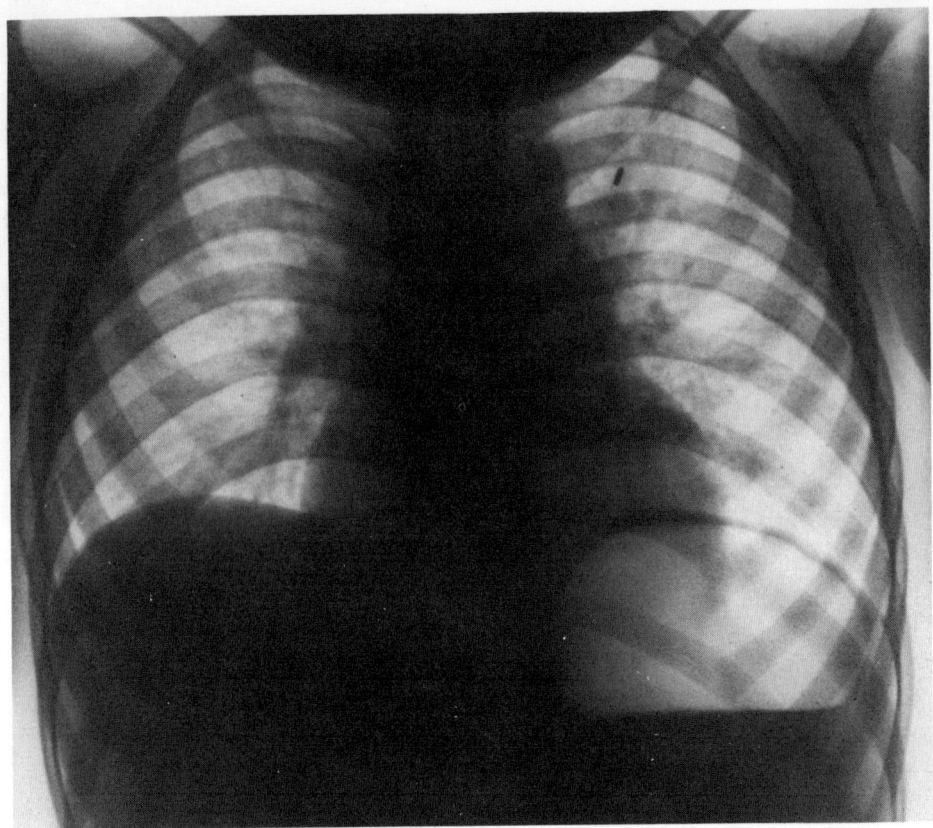

Fig. 1. Embolization of a Catheter to the left pulmonary artery.

Surgical removal of this foreign body out of caval vein, right heart or pulmonary artery carried a mortality of 2.3%; when the catheter was left in place death occurred in 39.6% of the cases. We, therefore, recommend the prompt removal of a broken catheter by cutdown of a peripheral vein and use of a Zeiss loop or by thoracotomy. Most catheter emboli occurred as a consequence of shearing the catheter with the sharp needle tip through which the catheter was inserted. We therefore developed a catheter which is introduced through a split canula and thus prevents this complication. By using a split-canula, the needle may be removed and sterile introduction of the catheter insured.

<u>Air embolism</u> is rare, it can be prevented by placing the patient in Trendelenburg position, while inserting the catheter through the external jugular or the subclavian vein and by securing the connection between venous catheter and infusion-set.

<u>Thrombosis</u> is a frequent complication of vena cava catheterization, and occurred clinically in 7% of our patients; at autopsy it was verified in 25% of our cases (Figs. 2 and 3).

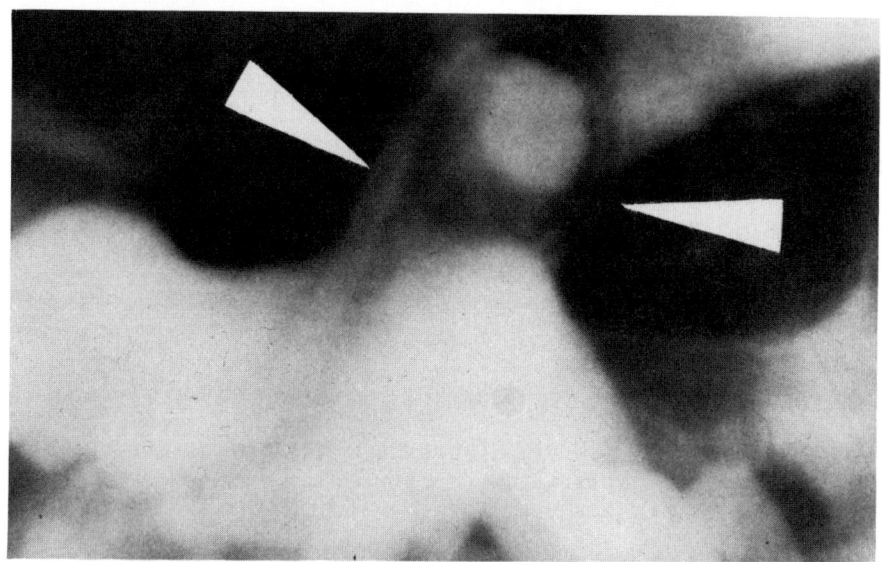

Fig. 2. Venogram of a patient with a subclavian - catheter. Partial occlusion of the vein by thrombosis.

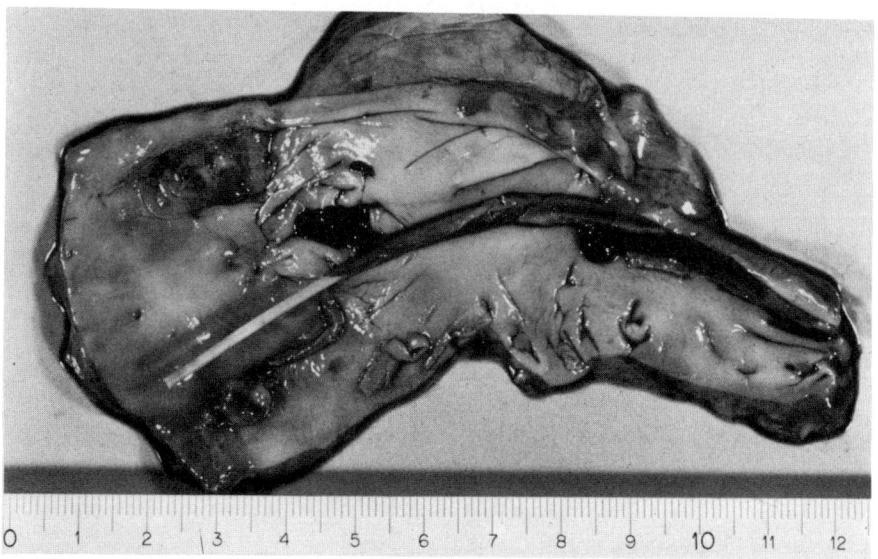

Fig. 3. Thrombus shearing a caval - catheter in the caval vein.

The occurence of thrombosis depends on local factors and is directly related to the length of time the catheter stayed in place. Therapeutic anticoagulation did not decrease the frequency of thrombosis.

Infections occurred clinically most frequently in the arm veins, and were found in 10% of the cases. On the basis of 4.000 bacteriological investigations the catheter tips were contaminated in 16.3%, skin swabs were positive in 22% and irrigation fluid from removed catheter contained bacteria in 11.5%. The frequency of bacterial contamination of a caval catheter depends on the primary disease, local infections, the technique of catheter placement and care and the time the catheter stayed in place. Therefore, catheters should not be used longer than absolutely necessary. If there is evidence of skin or vein irritation or fever of unknown origin the catheter should be removed immediately.

On the basis of postmortum examinations the catheter was not thought to have contributed to death in three of 373 cases autopsied. These three patients had basilic vein catheters, two of which were introduced by cutdown. One of these patients had extensive thrombosis of the vena cava and multiple pulmonary emboli, another septic thrombosis and the third had endocarditis involving the tricuspid valve.

A summary of the literature reveals that catheters introduced through a femoral vein carry a mortality of 4%, those through the basilic vein 0.4% and those inserted through to the subclavian vein 0.12%. No mortality was observed for catheters introduced through the external jugular vein. The inexperienced should use the approach via the basilic vein. This approach calls for meticulous care. An approach through tissue already damaged must be avoided.

Technique
1. - Aseptic technique and sterile introduction of the catheter
2. - Atraumatic venaepuncture
3. - Careful advancing of the catheter
4. - Radiological monitoring of catheter placement
5. - Do not retract the catheter while the needle is still in place to avoid catheter embolism.

Care
1. - Daily local antibiotic application at the catheter site
2. - Sterile dressing
3. - Careful fixation of the catheter to the skin
4. - Flush catheter daily
5. - Change infusion set
6. - Careful handling of the catheter connection to avoid air-embolism.

Based on the investigation of over 14.000 vena cava-catheters (literature and own study) we make the following recommendations concerning the use of catheters:

Indications for venous catherization
1. - Critical circulatory situation
2. - Preparation for an extensive surgical operation
3. - Surgical procedure on a high risk patient
4. - Parenteral nutrition
5. - Longtime infusions
6. - Administration of hypertonic solutions.

Approach
1. - Sites: external jugular, subclavian or basilic vein
   The approach via the lower extremities should be avoided.
   Puncture of the subclavian vein should be carried out by a physician experienced in this procedure.

Indications for catheter removal
1. - If catheter is no longer necessary
2. - Skin irritation or inflammation at the catheter site
3. - Venous thrombosis
4. - Pain
5. - Fever of unknown origin.

Reference
    Burri C., Gasser, D.   Vena cava - Catheter
                          Springer, Berlin-Heidelberg, New York, 1971

Address of the authors
    Prof. Dr. C. Burri and Dr. H.H. Passler, Department of Traumatology, D - 79 Ulm, Universitaetsklinik am Safranberg, Steinhovelstr. 9.

METABOLIC COMPLICATIONS OF TOTAL PARENTERAL NUTRITION

William C. Heird, Robert W. Winters and Stanley J. Dudrick

From the Department of Pediatrics, Columbia University College of Physicians & Surgeons, and Babies Hospital and the Department of Surgery, University of Texas School of Medicine at Houston, Houston, Texas

There is little question that total parenteral nutrition constitutes a valuable and often life-saving therapy for complex nutritional disorders of infants, children and adults. As the use of this technique has become widespread, the indications and contraindications for its use are beginning to emerge rather clearly. However, such widespread use of the technique has been attended by a sometimes alarming degree of abuse as reflected by the high rates of serious metabolic and septic complications which have been reported. It is the purpose of this paper to review the origin of various metabolic complications which may occur during the course of parenteral nutrition and, perhaps of greater importance, to present our views on how these complications can be minimized or avoided.

Disorders of Glucose and Electrolyte Metabolism

Table 1 shows a summary of the metabolic complications which have been reported in patients receiving total parenteral nutrition. A perusal of this table reveals that virtually every known aberration of sodium and water metabolism has occurred, viz, hyper- and hyponatremia, edema and dehydration. In the absence of complicating factors (e.g., failure to replace large losses of gastrointestinal secretions), hyponatremia and dehydration are nearly always secondary to the occurrence of serious hyperglycemia accompanied by profuse glycosuria and an attendant osmotic diuresis. Such a sequence involves a concurrent large sodium excretion which is roughly proportional to the intensity of the osmotic diuresis. Failure to replace these urinary losses of sodium and water leads to hyponatremia and dehydration.

Table 1: METABOLIC COMPLICATIONS OF TOTAL PARENTERAL NUTRITION

Glucose Disorders:	Hyperglycemia (Osmotic diuresis, hyperosmolar coma); hypoglycemia
Electrolyte Disorders:	Hyper- or hypo-natremia; -kalemia; -chloremia
Mineral Disorders:	Hyper- or hypo-calcemia; phosphatemia; magnesemia
Acid-base Disorders:	Hyperchloremic metabolic acidosis
N Metabolic Disorders:	Azotemia; Hyperammonemia; abnormal plasma aminograms
Vitamin Disorders:	Hyper- or hypo-vitaminosis
Fatty Acid Disorders:	Essential fatty acid deficiency
Trace Metal Disorders:	Zn, Cu deficiencies

Parenthetically, it should be noted that with significant hyperglycemia the plasma sodium concentration tends to fall. At high plasma levels, glucose behaves as a relatively impermeable solute. Therefore, its osmotic effect is confined to the extracellular fluid and elicits a shift of water from the intracellular to the extracellular fluid, thus diluting all constituents of the extracellular fluid of which sodium is quantitatively the most important.*

*This dilutional effect can be computed by calculating the plasma osmolality from the plasma glucose and sodium concentrations. Ordinarily, the plasma osmolality is closely approximated by the equation:

$$(mOsm) = 2 \times (Na^+)$$

Thus, in normal plasma having a sodium concentration of 140 mEq/l, the above equation gives a normal osmolality of 280 mOsm/kg of water (Winters, R. W., 1973). However, in the presence of hyperglycemia the mOsm contributed by glucose must be taken into account:

$$\text{mOsm from glucose} = \frac{\text{glucose in mg/100 ml} \times 10}{\text{molecular weight of glucose}} = \frac{mg/l}{180}$$

For example, the total osmolality of plasma having a glucose concentration of 540 mg/100 ml (or 5400 mg/l) and a sodium concentration of 125 mEq/l would be:

$$\frac{(mOsm)}{(mOsm/kg)} = 2 \times \frac{(Na^+)}{(mEq/l)} + \frac{(glucose)}{(mM/l)}$$

$$(mOsm) = (2 \times 125) + \frac{5400}{180} = 250 + 30 = 280 \text{ mOsm/kg water}$$

Failure to monitor the blood glucose carefully is one of the most frequent errors committed in the use of total parenteral nutrition. Since glucose is the sole source of caloric intake in this procedure, the daily load of glucose must be increased gradually during the initial period in order to allow the patient's endogenous mechanisms for glucose disposal to adapt to the large glucose loads presented. The possibility of an inadequate endogenous supply of insulin must also be borne in mind, particularly in older patients or premature infants as well as in patients with "latent" diabetes. The well-known glucose intolerance which so frequently accompanies stress, including the postoperative state, must also be borne in mind. The judicious use of insulin, with careful serial monitoring of blood sugar, is often helpful in minimizing glucose intolerance seen in the initial stages of total parenteral nutrition.

Hyperosmolar coma with extremely high concentrations of blood glucose (often accompanied by a concomitant hypernatremia) may occasionally be seen during the course of total parenteral nutrition, particularly if adequate monitoring is not carried out. This complication has been implicated in several fatalities of total parenteral nutrition.

Peripheral or pulmonary edema can occur if the sodium and water intake is excessive, particularly in the presence of such predisposing factors as compromised cardiac function, hypoproteninemia, etc.

Hypokalemia often occurs during the course of total parenteral nutrition. Initially, this situation probably results from the uptake of glucose by the liver (as in the patient with diabetic acidosis being treated with insulin and receiving no potassium intake). However, this effect is only transient. More sustanied hypokalemia usually occurs in the face of inadequate potassium intake and a concurrent brisk anabolic state, since for every gram of nitrogen deposited as new tissue, an average of 3 mEq of potassium is also laid down (Williams, G. S., et al., 1973). There are occasional adult patients who seem to develop hypokalemia while receiving as much as several hundred mEq of potassium per day. In such patients the anabolic demand for potassium can hardly account entirely for the hypokalemia; thus, other as yet poorly understood mechanisms must be operative.

Hyperkalemia may also be seen during the course of total parenteral nutrition. It is well known that acidosis tends to increase the plasma potassium concentration and that this defect involves both renal and extrarenal mechanisms (Williams, G. S., et al., 1973). This complication calls for a downward readjustment of potassium intake and/or correction of the acidosis. Again, serial monitoring of both plasma potassium and blood acid-base status are necessary for the early detection and correction of this type of abnormality.

Hypoglycemia may occur if total parenteral nutrition is suddenly stopped--e.g., due to dislodgement of the catheter. Presumably this complication occurs because endogenous insulin secretion (as well as other glucose disposing mechanisms), once adapted to the high glucose load, show a distinct lag in readjustment. It is for this reason that patients should be "weaned" from total parenteral nutrition to a continuous peripheral infusion of 5 or 10 percent glucose thereby allowing a gradual readjustment of the endogenous glucose-disposing mechanisms.

Disorders of Calcium, Phosphorus and Magnesium Metabolism

The most frequent abnormality of mineral metabolism seen in patients receiving total parenteral nutrition is hypophosphatemia. Predominantly, this abnormality occurs during the anabolic state. As in the case with potassium, the deposition of nitrogen in newly synthesized tissue is accompanied by a concurrent deposition of inorganic phosphorus per gm of nitrogen (Williams, G. S., et al., 1973). In addition, there may be a renal factor involved, at least initially. In some instances an initial inappropriate degree of phosphaturia occurs during the development of hypophosphatemia--i.e., a continuing renal excretion of phosphate despite plasma phosphate levels at which normal individuals would conserve phosphate completely (Heird, W. C., in press). This inappropriate phosphaturia is transient; ultimately, as the plasma phosphate level falls even lower, phosphate disappears almost completely from the urine. This transient inappropriate excretion of phosphate exaggerates the development of hypophosphatemia. One interpretation of this phenomenon is that certain amino acids entering the glomerular filtrate inhibit phosphate reabsorption until the filtered load of phosphate falls to levels at which it can be entirely reabsorbed despite the inhibition.

Hypophosphatemia may be accompanied by clinical signs; often, however, these signs are neither present nor specific. One important biochemical consequence of hypophosphatemia is the alteration which occurs in 2,3-diphosphoglycerate and other organic phosphate compounds of the erythrocyte (Travis, S. F., et al., 1973). These substances are known to have a significant influence upon the oxygen dissociation curve of hemoglobin. Thus, alterations induced by hypophosphatemia cause the oxygen dissociation curve to be shifted to the left, thereby reducing the amount of oxygen delivered to the tissues.

In view of the above considerations, careful attention must be paid to the amount of inorganic phosphate in the infusate. It is especially noteworthy that casein hydrolysates contain a significant amount of phosphate. The crystalline amino acid mixtures, however, do not contain phosphate. Therefore, when the nitrogen source is changed from a casein hydrolysate (e.g., Amigen^R) to a crystalline amino acid mixture (e.g., FreAmine^R), more inorganic phosphate must be added.

Hyperphosphatemia during the course of total parenteral nutrition is uncommon. When it occurs, it usually signifies an excessive intake of phosphate--assuming, of course, that the patient is not azotemic. Similarly, hypercalcemia occurs with excessive calcium intake but may also occur if excessive Vitamin D is added. In infants, hypercalcemia consistently accompanies hypophosphatemia (due to an inadequate phosphate content of the infusate); this situation occurs even in the absence of a concurrent intake of calcium (Heird, W. C., in press).

Hypocalcemia, even in the absence of the calcium intake, is uncommon in patients receiving total parenteral nutrition except, of course, under unusual circumstances. One such circumstance is correction of hypophosphatemia by administration of a large amount of exogenous phosphate in the absence of a supplementary calcium intake. In this situation, calcium may fall rapidly and tetany may result.

Hypomagnesemia has been recorded in a few instances. The requirement for magnesium seems to be greater when crystalline amino acid mixtures, rather than protein hydrolysates (particularly casein hydrolysates), are used as the nitrogen source. The reason for this situation is unknown.

As in the case with abnormalities of glucose and electrolyte metabolism, all of the above disorders of mineral metabolism can be avoided if the physician is fully aware of the specific mineral composition of the infusate and if a careful monitoring schedule for plasma total calcium, inorganic phosphate and magnesium is adhered to.

Acid-Base Disorders

Patients receiving either casein hydrolysates or fibrin hydrolysate as the nitrogen source of total parenteral nutrition rarely develop acid-base disorders. If they do, there is usually an obvious etiology for the disorder. However, with the advent of crystalline amino acid mixtures, especially NeoAminosol^R (an experimental preparation patterned after the free amino acid composition of Aminosol^R) and to a lesser extent FreAmine^R, a number of patients, especially infants and young children, developed a hyperchloremic metabolic acidosis. In the infants, the acidosis often was of significant proportions.

The mechanism of this acidosis has now been elucidated (Heird, W. C., et al., 1972). All solutions used as nitrogen sources for total parenteral nutrition have acid pH values (between pH 5 and 6) and some degree of titratable acidity. However, neither of these factors are physiologically relevant. Rather the most important factor, quantitatively, is the presence of cationic amino acids (arginine and lysine). In the crystalline amino acid mixtures, these cationic amino acids are present as the hydrochloride salts; in the hydrolysate preparations, on the other hand, the positive charges of these amino acids are matched by anionic amino acids and peptides. Metabolism of either arginine or lysine generates hydrogen ion which, in the case of the crystalline amino acid mixtures, is tantamount to infusion of hydrochloric acid. With the hydrolysates, however, the hydrogen ion generated by metabolism of the cationic amino acids is matched by an equivalent amount of bicarbonate generated from metabolism of the anionic amino acids and peptides. Thus, a net acid load results in the former case but not in the latter.

The acidosis occurring secondary to infusion of crystalline amino acid mixtures need not be serious. Once it is recognized, it can readily be treated with sodium bicarbonate or by the addition of lactate or acetate to the infusate. Reformulation of the amino acid mixtures so as to avoid using the hydrochloride salts will definitively solve this problem. This modification is technically feasible and hopefully will be instituted by the manufacturers.

Hyperammonemia

It has been recognized for many years that both casein and fibrin hydrolysates contain relatively high concentrations of free ammonia (30 to 50 mg/100 ml). This preformed ammonia can be readily metabolized by adults without liver disease. However, infants receiving either of these hydrolysates develop modest elevations of blood ammonia (Johnson, J. D., et al., 1972). Usually this modest hyperammonemia is unaccompanied by symptoms, although there are a few reports of severe symptomatic hyperammonemia occurring in infants receiving large loads of hydrolysates (Ghadimi, H., et al., 1971; Walker, F. A., 1971).

FreAmine[R], the crystalline amino acid mixture currently available in the United States, contains negligible quantities of free ammonia. Theoretically, therefore, it should be advantageous in minimizing the risks of hyperammonemia. Thus, it was surprising to discover that some infants (Heird, W. C., et al., 1972) as well as adults (Dudrick, S. J., et al., 1972) receiving FreAmine[R] developed significant hyperammonemia. In the case of the infants, the blood levels of ammonia exceeded 800 mcg/100 ml and were assciated with convulsions. Further study of this phenomenon has revelaed that this hyperammonemia can be treated by arginine hydrochloride, arginine glutamate or ornithine hydrochloride and can be prevented by supplementation of the amino acid mixture with arginine (Heird, W. C., et al., 1972). Thus, it would appear that this particular mixture (i.e., FreAmine^R) is relatively deficient in arginine, particularly when the period of total parenteral nutrition is prolonged. Two lines of evidence support this contention: (a) plasma levels or arginine and/or ornithine tend to be low in these hyperammonemic infants, and the urinary excretion of urea falls dramatically during development of hyperammoniemia.

Abnormalities of Hepatic Function

One of the most confusing aspects of total parenteral nutrition involves its effect on hepatic function. A number of patients develop elevations of serum transaminases (SGOT and SGPT) during the course of total parenteral nutrition. In most instances these elevations are transient and are not accompanied by other signs of deranged hepatic function (Heird, W. C., et al., 1972). However, Cohen has encountered a number of patients receiving total intravenous alimentation who develop both hepatomegaly and hepatic tenderness in conjunction with elevations in serum transaminase levels (Cohen, M. I., et al., 1973). These changes were sometimes accompanied by elevated levels of serum bilirubin. In addition, Cohen has shown that the damage to liver explants maintained in the presence of the fibrin hydrolysate, Aminosol^R, is equivalent to that of explants maintained in the presence of carbon tetrachloride, a known toxin (Cohen, M. I., personal communication).

At autopsy, premature infants who died from non-hepatic causes while receiving intravenous alimentation were found to have enlarged livers which were characterized histologically, by fibrosis, fatty infiltration, and bile stasis (Driscoll, J. M., et al., 1972).

It is clear that the effect of total parenteral nutrition on hepatic function is poorly understood. However, to our knowledge, no permanent hepatic damage has been reported as a result of total parenteral nutrition.

Changes in Blood Urea Concentration

In the absence of underlying renal disease or disorders of hydration, blood urea concentration during total parenteral nutrition usually remains within normal limits. However, patients receiving higher intakes of nitrogen tend to have higher blood urea concentrations than those receiving lower total nitrogen intakes. Indeed,

Abnormalities in Plasma Amino Acid Patterns

in pediatric patients, reducing the total protein intake from 4 gm/kg/d to 2.5 gm/kg/d has little effect on nitrogen balance or rate of weight gain, but tends to significantly decrease the blood urea concentration (Heird, W. C., unpublished results).

Aberrations in the plasma aminograms of patients receiving total parenteral nutrition has been seen by a number of investigators. In general, the aminogram tends to mimic the amino acid composition of the infusate (Stegink, L. D., et al., 1971; Heird, W. C., et al., unpublished results). The significance of these aberrations of plasma amino acids is not clear. They are probably of more concern in infants, in whom development and maturation of the central nervous system is not complete, than in adults whose brains are fully developed.

High plasma levels of glycine are seen in most infants receiving total parenteral nutrition using crystalline amino acid mixtures in which the non-essential nitrogen is provided largely by this inexpensive amino acid. In view of the fact that some inborn metabolic errors characterized by hyperglycinemia are accompanied by mental retardation, this hyperglycinemia may be important. In these disorders, the causal relationship between hyperglycinemia and mental retardation is unclear; nonetheless, the association of the two is of some concern for those who care for infants receiving total parenteral nutrition.

High plasma levels of methionine have also been noted in patients receiving the crystalline amino acid mixture FreAmine^R. Since the methionine content of this mixture consists of both the D- and L- forms, it is possible that the high levels represent only the D- form. This problem can be resolved by using only the L- form of methionine in the manufacturing process.

In some premature as well as term infants cystine may be an essential amino acid (Sturman, J. A., et al., 1970). Infants for whom this is true lack one of the enzymes responsible for the conversion of methionine to cystine (i.e., cystathionase). The exact number of infants having this defect is not known, but it is thought to be significant. Therefore, it would seem to be desirable to add cystine to the infusate. Because of its low solubility, however, this is not possible. Thus alternate ways of providing cystine (perhaps as a soluble metabolizable compound) must be explored.

The problem of adding cystine may not be limited strictly to the pediatric patient. Stegink has shown that normal adult subjects receiving Amigen^R by nasogastric tube maintain normal plasma cystine levels whereas the same subjects, when given this nitrogen source intravenously, consistently develop low plasma levels of this amino acid (Stegink, L. D., personal communication). One possible implication of these studies is that the route of administration plays a role in whether or not an amino acid is "essential".

Essential Fatty Acid Deficiency

Total parenteral nutrition as it is currently practiced in the United States employs infusates which are entirely devoid of lipid. Thus, it is not surprising that clinical essential fatty acid deficiency has been encountered. This deficiency state is characterized by a "scaly" skin rash. It can be treated by infusions of Intralipid^R, an intravenous fat preparation rich in unsaturated fatty acids (Caldwell, M. D., et al., 1972). In contrast to the relative rarity of clinical essential fatty acid deficiency, chemical evidences of this deficiency occur with high frequency (Dudrick, S. J., in press). The typical chemical changes seen in the plasma are a decrease in linoleic acid with a reciprocal increase in Δ 5,8,11-eicosatrienoic acid. This general area requires much more study, especially in the pediatric patient. Essential fatty acid deficiencies may affect the essential fatty acid status of the brain lipids, and such alterations may be of great importance in determining the ultimate function of the developing nervous system of the premature or full term infant.

Other Complications

A variety of miscellaneous complications have been reported and more will likely be uncovered as they are searched for. For example, isolated examples of copper deficiency as well as zinc deficiency have been seen. Bleeding, due to failure to provide Vitamin K, has also occurred. The multivitamin preparation commonly used in total parenteral nutrition (MVI^R) contains no Vitamin K. This preparation also lacks folic acid and Vitamin B$_{12}$. Thus, these vitamins must also be provided if deficiency states are to be avoided.

The optimal requirements for both trace minerals and vitamins during total parenteral nutrition are subjects which demand much systematic study.

Detection and Prevention of Metabolic Complications

There are several major steps which must be taken to detect and/or to prevent metabolic complications during total parenteral nutrition.

First, the individual needs of each patient for each nutrient (i.e., glucose, nitrogen, electrolytes, vitamins and minerals) must be taken into account. This step is particularly important for those patients who are metabolically unstable with day to day variations in their requirements for glucose and electrolytes.

Second, in all patients, the glucose concentration of the infusate must be increased slowly--sometimes, over a period of several days--in order to allow "adaptation" of the body's glucose-disposing mechanisms. The rate of increase of glucose should be highly individualized and should be dictated by feedback information derived from the measurements of blood and urinary sugar.

Third, adherence to a strict schedule of chemical monitoring of all relevant variables is essential. Table 2 shows the monitoring schedule which the Babies Hospital group has developed.

Table 2: MONITORING SCHEDULE FOR TOTAL INTRAVENOUS ALIMENTATION USED AT BABIES HOSPITAL

Chemical Variable	Initial Period	Later Period
Plasma Na^+, K^+, Cl^-, BUN & glucose	Daily	3 times/week
Blood acid-base status	Daily	3 times/week
Plasma Ca^{++}, P and Mg^{++}	3 times/week	1-2 times/week
Blood NH_3	3 times/week	3 times/week
Liver function tests	2 times/week	Weekly
Hemogram	2 times/week	2 times/week
Protein Electrophoresis	Weekly	Weekly
Urinary glucose	4 times/week	Daily

Although this schedule is designed for infants, it is equally applicable to adults. The schedule is divided into two phases, an initial one during which the patient is adapting to the increasing glucose intake and a later one when he has achieved a more stable state metabolically. However, in the face of intermittent stress (i.e., surgery, infection, etc.), the schedule outlined under the initial period should be adopted since metabolic complications under these conditions are likely to be more frequent.

Table 3 summarizes the results of adherence to this strict monitoring schedule in a group of 21 infants and children with a variety of surgical disorders who received total parenteral nutrition for a total of 777 patient-days.

Table 3: INCIDENCE OF CHEMICAL ABNORMALITIES DURING TOTAL INTRAVENOUS ALIMENTATION

Constituent	High	Percent of all Determinations Normal	Low
Sodium	2%	91%	7%
Potassium	13%	77%	10%
Calcium	2%	89%	9%
Phosphorus	46%	48%	6%
Magnesium	27%	72%	1%
Urea	8%	44%	48%
Glucose	22%	62%	16%
Base Excess	5%	74%	21%
SGOT	58 (38*)%	42%	--
SGPT	36 (25*)%	64%	--

*With liver disease.

The data in this table demonstrate that with few exceptions the concentration of the plasma sodium, potassium, total calcium, urea, and glucose can be maintained within the normal limits if the patient is monitored closely and the composition of infusate adjusted in accordance with appropriate feedback information. In the case of magnesium and inorganic phosphate, the significant number of elevated values probably reflect an overestimate of requirements and a downward adjustment of these requirements seem indicated. In the case of blood base excess, the relatively high incidence of low values indicates the frequency with which metabolic acidosis occurred. This disorder was especially common in patients receiving NeoAminosol^R but occurred, although less frequently and with less severity, in those receiving FreAmine^R, as well. The high incidence of elevated serum transaminases is explained partially by the high incidence of liver disease seen in these patients. However, as discussed earlier, elevations of these enzymes also occur in patients without liver disease but in these cases the rises are usually transien

Conclusion

Total parenteral nutrition is fraught with a seemingly endless variety of metabolic complications. Most of these can be recognized early and corrected by appropriate adjustment of the composition of the infusate. To maximize the positive effects of this often life-saving technique while minimizing the risks requires a "team approach". The necessary team should be headed by a physician, but he must have strong support from responsive chemistry and microbiology laboratories as well as from pharmacy and nursing services (Heird, W. C., et al., 1972). It is the firm belief of the authors that institutions unable to mount such a team should refer patients in need of total parenteral nutrition to institutions who have organized a team and who have a proven record of confidence.

Acknowledgements

The original work of these authors is supported by a grant from the National Institute of Child Health & Human Development (HD-03993).

Dr. Winters is Career Scientist of the Health Research Council of the City of New York (I-309).

REFERENCES

Caldwell, M. D., Jounson, H. T. and Othersew, H. B., J Pediat., 81:233, 1972.

Cohen, M. I., Pediatr. Res., 1:334, 1973.

Cohen, M. I., Personal Communication.

Driscoll, J. M., Jr., Heird, W. C., Schullinger, J. N., et al., J Pediatr. 81:145, 1972.

Dudrick, S. J., in: Intravenous Alimentation of High Risk Infants, Gov't Printing Office, In Press.

Dudrick, S. J., MacFayden, B. V., Jr., VanBuren, C. T., et al., Ann Surg., 176:259, 1972.

Ghadimi, H., Abaci, F., Kumar, S, and Rathi, M, Pediatrics 48:995, 1971.

Heird, W. C., in: Intravenous Alimentation in High Risk Infants, Gov't Printing Office, In press.

Heird, W. C., Unpublished results.

Heird, W. C., Dell, R. B., Driscoll, J. M., Jr., et al., New Engl. J. Med., 287:943, 1972.

Heird, W. C., Driscoll, J. M., Schullinger, J. N., et al., J. Pediatr., 80:351, 1972.

Heird, W. C., Nicholson, J. F., Driscoll, J. M., Jr., et al., J. Pediatr. 81:162, 1972.

Heird, W. C. and Nicholson, J. F.: Unpublished results.

Johnson, J. D., Albritton, W. L. and Sunshine, P., J. Pediatr., 81:154, 1972.

Stegink, L. D., Personal Communication.

Stegink, L. D. and Baker, G. L., J. Pediatr., 78:595, 1971.

Sturman, J. A., Gaull, J. G. and Raiha, N. C. R., Science, 169: 74, 1970.

Travis, S. F., Sugerman, H. J., Ruberg, R. L., et al., New Engl. J. Med., 285:763, 1973.

Walker, F. A., New Engl, J. Med., 285:1324, 1971.

Williams, G. S., Klenk, E. L. and Winters, R. W., in: The Body Fluids in Pediatrics, Winters, R. W., Ed., Little, Brown and Co., Boston, 1973.

Winters, R. W., in: The Body Fluids in Pediatrics, Winters, R. W., Ed., Little, Brown and Co., Boston, 1973.

INFECTION IN ASSOCIATION WITH INTRAVENOUS FEEDING

Murray F. Brennan

Department of Surgery of Harvard Medical School, Peter Bent Brigham Hospital, Boston, Massachusetts

The advent of a method of intravenous feeding that allows growth, development and restoration of body deficits has been a major advance in surgical care (23). Like most advances, the initial, and on occasions, unbridled enthusiasm of some of the disciples has resulted in a rebound of criticism and re-evaluation. We believe that this phase is now passing and total parenteral nutrition is assuming its rightful place as a major surgical adjunct. With many surgical units in this country now involved in meaningful research and development, further exegesis can be expected. The problem of infection in association with this therapy has received much attention (3,6,10,12,18,48). However, many of the problems with infection have arisen as a consequence of ignoring both the recommendations of the innovators (51) and standard aseptic procedures.

The problem is really twofold:
a) the "general" problem, i.e., the infected seriously ill patient, and
b) the "specific" problem, i.e., that related to (1) the presence of an indwelling intravenous line and (2) the nature of the nutrient mixture being delivered via this circulatory portal.

The infected patient, whether locally infected at the site of introduction of the intravenous catheter, as in a burn, or with disseminated sepsis, as in diverticulitis, is, in theory, at added risk from septicemia. In these patients there is a marked increase in the possible sources of line contamination. This particular aspect has not markedly increased our own incidence of line-related septicemia although where a catheter is positive, that organism can frequently be found at another site remote from the site of entry of the indwelling line (9). The other important environmental sources are the personnel and the room contaminants; in our hospital the cleanest air available contains 1 to 5 organisms per cubic foot with the use of ultraviolet doorway light screening.

The severely ill, infected patient frequently has the added problem of receiving multiple antibiotics, steroids, and often immunosuppressive drugs, all acting to diminish resistance and allow pathogenic overgrowth of the normal flora of skin and mucosal tracts. All these factors combine to provide an excellent environment for fungal and bacterial growth, and we have noted an apparent increase in the general incidence of disseminated candidiasis in patients coming to autopsy (Table 1).

Table 1: Incidence of Candidiasis at Autopsy
PBBH, March 15 to June 15*

	1972			1971
Total autopsies	88			87
Fungal infection	9			7**
Candida	6			6
Mucor	3			-
Aspergillosis	-			1
Predisposing factors	- Hematological disease	6/13		6
	- Transplant	1/4		
	- Burns	1		
	- Postoperative	1		

*A.K.ABBAS
CPC, PBBH, June 26, 1970
**Four (4) restricted to gastrointestinal tract.

The observation that the serum inhibition of Candida albicans seen in normal patients (38) is lost in patients with Candida septicemia (12) raises interesting possibilities. In contradistinction to this we have found that patients can develop positive precipitin tests from candida (47) without clinical evidence of candidemia (blood culture negative, no fever)(16). These patients are receiving the "amphotericin flush" (15) (see below), and this implies that catheter-related candidemia may be dependent on repetitive inoculation to produce clinical candidemia. Finally, there is the cyclical variation of neutrophil bactericidal capacity (1,2) without change in the ability to phagocytose, which is additionally impaired by corticosteroids. Perhaps the temporal inoculation of organisms is all-important in the development of clinical bloodstream infection. The specific problem of the indwelling catheter and the delivered nutritive solutions is, however, what we are primarily concerned with here. The catheter complications have already been discussed in some depth (8,9,13,17,18,20,22,25,28,44, 45,49,50,53).

Table 2: Intravenous Catheterization: Infectious Complications

Druskin 1963	0% positive at less than 48 hours
	52% positive at greater than 48 hours
Collins	7% positive at 24 hours
	27% positive at 96 hours
Freeman 1972	100% positive at 96 hours

We have been predominantly concerned and interested in the problem of candidemia in association with this mode of therapy, as have others (18). However, the majority of comments apply equally well to bacterial septicemias. Certainly the regular catheter used intravenously is commonly positive at the time of withdrawal and culture (Table 2). The incidence of bacteriologically positive catheters increases when additives are used with the more standard solutions (14). This is also true of the incidence of candidemia and intravenous feeding (Table 3).

Table 3: Candidemia and Intravenous Feeding: PBBH - 1970-1971

Route	Duration of Line (Days)	Catheter	Blood
1. Subclavian	72	+	+
2. Subclavian	67	-	+
3. Subclavian	15	+	+
4. Subclarian	44	+	+
5. Subclavian	9	+	+
6. A.V. Shunt	133	+	+
7. A.V. Shunt	10	+	+
8. Subclavian	3	-	+
	44± 16	6/8	8/8

The physical presence of a foreign body in the circulation, besides providing a portal of entry, provides a nidus for thrombus formation and bacterial or fungal growth. The latter was demonstrated in 1951 (24) and confirmed recently (4).

The association of duration of catheterization, with the incidence of catheter positive sepsis has been widely debated. Some believe duration to be a factor (14,18,22,31), and others do not (10,44). The latter group believe that the catheter is contaminated at the time of introduction. In our experience (Table 4) duration does appear to be a factor. The ability of the nutritive solutions to support bacterial and fungal growth is now well established (12, 14,43). Candida albicans will grow particularly well in casein hydrolysates with dextrose, and in intravenous soy bean fat emulsions.

Table 4: Intravenous Feeding: Infectious Complications*

Mean duration of line without septicemia	= 16 ± 4
Mean duration all catheters	= 29 ± 9
Mean duration for candidemia	= 44 ± 16
Mean duration for bacterial septicemia	= 24 ± 6

*PBBH 1970-71 (days, mean $\pm$ S.E.M.)

The differential between growth in casin and fibrin hydrolysates raises interesting speculation. Is it the contained amino acids which are important? Does the high content of glutamic acid (in the casein hydrolysate), a known growth substrate for candida (34) make the difference, or, as with bacteria, does the inhibition in the fibrin hydrolysates by the content of 1) methionine (36) or 2) glycine (33) make the difference? The latter seems unlikely in view of the ability of candida to grow in glycine (35). How is it that candida can survive without a nitrogen source as in the fat emulsion? This would be no surprise in the case of bacteria (43), E. coli is able to grow in distilled water by extracting ammonia and carbon dioxide from the air (11). Perhaps we are missing the point and it is the essential requirements of minerals that is important. Certainly growth can be enhanced by phosphate (34). These and other growth requirements are currently under investigation by Dr. Ruth Kundsin in our Surgical Bacteriological Laboratories.

Apropos of the problem is the use of nitrogen assimilation patterns to identify various yeast species (52). This has recently been taken further by Lee (37), whose work raises the interesting conjectural possibility that we may be able to not only identify yeast species by their ability to hydrolyze particular amino acids, but to provide specific organism inhibitors by the addition of specific components. While this approach would be clinically relevant, it is fraught with danger; substances that might inhibit fungal or bacterial growth may well have deleterious effects on man, e.g., the toxic effects of high doses of glycine (21,32,40). Such an approach would be additionally hazardous in the presence of any degree of impairment of hepatic function. All these questions are purely rhetorical when we consider that any organism or fungus once lodged within the catheter lumen has access to all nutritive requirements for growth, and bloodstream infection is inevitable. Clinical septicemia is then solely dependent on the integrity of the intravascular defense mechanisms (39).

The Prevention of Line-Related Sepsis

As is true with so many facets of medical and surgical care, awareness of the problem is often accompanied by progressive and rewarding change. We believe that with careful attention to detail, the sepsis problem can be reduced to an acceptable level in these seriously ill patients. The major points that we consider to be important are:

1) Patient Selection: Intravenous feeding with multiple substrates is not a panegyric for all ills, and is not indicated in routine postoperative fluid maintenance. The indications for use of this therapeutic adjunct are becoming more clear.

2) Line Placement: The diligent attention to aseptic line placement remains paramount and, where practical, requires the presence of an experienced physician in an effort to decrease the complications of placement (10).

3) Catheter Type: The type of catheter being used is receiving much greater attention. The silicone catheter*, which is flexible, and does not require the persistence of a sharp bevelled needle around the outside, appears to be an advance. Our own trials with this catheter are promising. Debate has continued about the importance of the dimensions and length of the indwelling catheter; some (50) believe it to be influential while others do not (10). It is uncommon to see infection with the peripherally placed "scalp vein"** needle (50). This reinforces the need to be able to use peripherally administered nutritive solutions such as soy bean oil.

4) We routinely cleanse the skin with benzalkonium hydrochloride*** and change the povidone-iodine**** dressing over the

---

*Medicath (Cheeseborough)
**Butterfly (Abbott)
***Zephiran (Winthrop)
****Betadine (Purdue-Frederick)

puncture site along with the intravenous tubing twice a week. The entry wound site is cultured at the sign of any erythema. With any fever, it is our practice to take a "drawback" blood culture.

5) Specific Prevention of Candida Sepsis: Following a recurring problem (Tables 3 and 5) of candidemia in patients on long term intravenous feeding, we developed and investigated, both in vitro and in vivo, the "amphotericin flush." (15) This has virtually eradicated catheter-induced candidemia, and at the dose that we employ toxicity has not been manifest. The exact role of the "amphotericin flush" as a routine measure is not yet clearly defined. The on-going controlled clinical trial of Dr. Filler, Dr. Goldman, Dr. Gardner, and colleagues at the Children's Hospital Medical Center in Boston should provide further elucidation. At this time we feel that the "amphotericin flush" will take its rightful place in those patients who are placed on long term feeding, and it may be that we shall come to some arbitrarily defined limit (e.g., twenty days) at which time the "amphotericin flush" will be introduced. Currently all our patients receive this therapy from the initiation of intravenous nutritive feeding.

Table 5: Candidemia and Intravenous Feeding

Authors	Incidence	Organism	Catheter Positive	Blood Positive
Boeckman, et al. (1970)	6/15 40%	C. albicans	5/6	2/6
Aschcraft, et al. (1970)	5/22 23%	C. albicans C. guilliermondi	2/5	5/22
Curry, et al. (1971)	(a) 8/49 16% (b) 22/33*	C. albicans T. glabrata C. paratrusei	- 12/22	8/8 33/33*
PBBH (1970/71)	4/28** 14%	C. albicans	2/4	4/4
Bernard, et al. (1970)	7/27***	C. tropicalis C. albicans	7/27	3/27

*All fungal septicemias, 22 on IV feeding.
**Further serological positive.
***All 27 subclavian lines, not necessarily for IV feeding.

6) The possibilities of modifying the contents of such mixtures so as to contain a bacterial inhibitor has been alluded to. We believe it will be from examination of the contents of those solutions that do not support microorganisms that this information will come.

7) Contamination of the solutions at the time of preparation is a major problem, and it is our practice for a trained pharmacist to prepare all these solutions within the pharmacy under laminar flow hood conditions. The actual design of the delivery container is important as was so ably illustrated by the study on the screw top bottle and delivery set (25). The plastic vacuum delivery bags are theoretically less likely to be contaminated. This has been confirmed by us (35).

8) Storage of the Prepared Mixtures: Solutions, once prepared are stored in the cold and discarded after thirty-six hours. Storage in the cold can insure inhibition, but not eradication of fungi (14). Others (29) have more prolonged storage times without apparent increased incidence of sepsis, and discard mixtures on the appearance of any turbidity.

9) Early Diagnosis of Catheter/Solution Sepsis: The routine use of serological tests where available has already been mentioned.

The Nitroblue tetrazolium test (NBT), as applied to the infective problems (30) presents an intriguing development. This phenomenon of reduction of colorless Nitroblue tetrazolium to blue/black forazan by the neutrophil of normal man in vitro (7) was applied to the phagocyte in vivo during bacterial infection in infants by Park in 1968 (46). Later confirmation was obtained with adults (27,41) in bacterial infection and in fungal infection (5,46).

No value can be attached to this test under the age of two months (26) and corticosteroids can produce false negatives (42). A greater value of NBT positive neutrophils than 11 per cent has been accepted as positive (30,31,41,42,46). The application of this test to indwelling catheter related infections (31) requires that other causes of infection be excluded, which in our seriously ill patients would limit the usefulness of the test. However, in such patients it would be comparable to the evaluation of an unexplained fever. Like fever, the control of the infection results in a return to normal of the NBT test (30,41). The speed with which the test can be performed is a further factor in favor of its use.

SUMMARY

Total parenteral nutrition is a lifesaving adjunct in surgical care. Diligent attention to detail, insures that infection in association with this therapy, is kept at acceptable levels.

ACKNOWLEDGEMENTS

The continual support and encouragement of Dr. Francis D. Moore and Dr. Ruth B. Kundsin is gratefully acknowledged. Work on which this presentation is based would not have been possible without the diligent attention to detail of Miss M.R. Ball, Miss M. MacBurney, Miss H. Bodman, and Dr. Stuart Jones.

REFERENCES

1. Alexander, J.W. Arch. Surg., 103:435, 1971.

2. Alexander, J.W., and Meakins, J.L. Ann. Surg. (in press).

3. Altman, R.P., and Randolph, J.G. Ann. Surg., 174:85, 1971.

4. Anderson, A.O., and Yardley, J.H. New Eng. J. Med., 286:108, 1972.

5. Anderson, B.R. Lancet, (ii): 317, 1971.

6. Ashcraft, K.W., and Leape, L.L. JAMA, 212:454, 1969.

7. Baehner, R.L., and Nathan, D.G. Science, 155:835, 1967.

8. Bansmer, G., Keith, D., and Tesluk, H. JAMA, 167:1606, 1958.

9. Bentley, D.M., and Lepper, M.H. JAMA,206:1749, 1968.

10. Bernard, R.W., and Stahl, W.M. Ann. Surg., 173:191, 1971.

11. Bigger, J.W., and Nelson, J.H. J. Path. Bact., 55:321, 1943.

12. Boeckman, C.R., and Krill, C.E. J. Pediatr. Surg., 5:117, 1970.

13. Bogen, J.E. Surg. Gynec. & Obstet., 110:112, 1960.

14. Brennan, M F., O'Connell, R.C., Rosol, J., and Kundsin, R.B. Arch. Surg., 103:705, 1971.

15. Brennan, M.F., Goldman, M.H., O'Connell, R.C., Kundsin, R.B., and Moore, F.D. Ann. Surg., 1972 (in press).

16. Brennan, M.F., Jones, S.A. Kundsin, R.B., and Moore, F.D. (to be published).

17. Burri,

18. Curry, C.R., Quie, P.G. New Eng. J. Med., **285**:1221, 1971.

19. Collins, R.N., Braun, P.A., Zinner, S.H., and Kass, E.H. New Eng. J. Med., **279**:340, 1968.

20. Dennis, D.C., Peterson, C.G., and Fletcher, W.S. J Trauma, **8**:177, 1968.

21. Doolan, P.D., Harper, H.A., Hutchin, M.E., and Alpen, E.L., J. Clin. Invest., **35**:888, 1956.

22. Druskin, M.S., Siegel, P.D. JAMA, **185**:966, 1965.

23. Dudrick, S.J., Wilmore, D.W., Vars, H.M., and Rhoads, J.E. Surgery, **64**:134, 1968.

24. Duhig, J.V., and Mead, M. Med. J. Aust., **1**:179, 1951.

25. Duma, R.J., Warner, J.F., and Dalton, H.P. New Eng. J. Med., **284**:257, 1971.

26. Editorial. Lancet, (**ii**):909, 1971.

27. Feigin, R.D., Shackelford, P.G., Choi, S.C., Flake, K.K., Franklin, F.A., and Eisenberg, C.S. Pediatrics, **78**:230, 1971.

28. Fekety, F.R., Thoburn, R. Hopkins Med. J., **121**:133, 1967.

29. Filler, R.M. Hospital Practice, **7**:79, 1972.

30. Freeman, R., and King, B. Lancet, (**ii**):1154, 1971.

31. Freemen, R., and King, B. Lancet, (**i**):992, 1972.

32. Harper, H.A., Najarian, J.A., and Silen, W. Proc. Soc. Exper. Biol. Med., **92**:558, 1956.

33. Joh, R.V., and Russell, A.D. J. Pharm. and Pharmacol., **15**:346, 1963.

34. Johnson, S.A.M., Guzman, M.G., and Acquilera, C.T. Arch. Derm. Syph., **70**:49, 1954.

35. Kundsin, R.B. (personal communication).

36. Lark, C., and Lark, K.G. Biochim. Biophys. Acta, **49**:308, 1961.

37. Lee, K.L., Watson, P.R., and Reca, M.E. Ann. Meeting Abstracts, 1972.

38. Louria, D.B., and Brayton, R.G. Nature (London), 201:308, 1964.

39. Lupin, A.M., Dascomb, H.E., Seabury, J.H., and McGinn, M. Antimicrobial Agents and Chemotherapy, 1:10, 1961.

40. Malvey, P., Rousseau, C., and Cardon, J. Presse Med., 69: 917, 1961.

41. Matula, G., and Paterson, P.Y. New Eng. J. Med., 285:311, 1971.

42. Matula, G., and Paterson, P.Y. Lancet, (i):803, 1971.

43. Michaels, L., and Ruebner, B. Lancet, (i):772, 1953.

44. Moran, J.M., Atwood, R.P., and Rowe, M.I. New Eng. J. Med., 272:554, 1965.

45. Norden, C.W. J. Infect. Dis., 120:611, 1969.

46. Park, B.H., Fikrig, S.M., and Smithwick, E.M. Lancet, (ii): 532, 1968.

47. Preisler, H.D., Hasenclever, H.F., and Henderson, E.S. Amer. J. Med., 51:352, 1971.

48. Prochazka, J.V., Lucas, R.N., Beauchamp, C.J., and Strauss, R.G. Amer. J. Dis. Child., 122:255, 1971.

49. Sachs, F.L. New Eng. J. Med., 277:433, 1967.

50. Smits, H., Freeman, L.R. New Eng. J. Med., 276:1229, 1967.

51. Wilmore, D.W., and Dudrick, S.J. Arch. Surg., 98:256, 1969.

52. Van Uden, N., Vidal-Leiria, M., and Buckley H.R. Ann. Soc. Belg. Med. Trop., 44:619, 1964.

53. Zinner, S.H., Denny-Brown, B.C., Braun, P., Burke, J.P., Toala, P., and Kass, E.H. J. Infect. Dis., 120:616, 1969.

CONTRIBUTORS

Antoon, A.	Department of Pediatrics, Harvard Medical School, Massachusetts General Hospital and Shriners Burns Institute, Boston, Massachusetts
Biebuyck, J.F.	Department of Anesthesiology, Harvard Medical School, Massachusetts General Hospital and Shriners Burns Institute, Boston, Massachusetts
Bjordal, R.	Department of Pediatric Surgery, Rikshospitalet, Oslo, Norway
Bode, H. H.	Department of Pediatrics, Harvard Medical School, Massachusetts General Hospital and Shriners Burns Institute, Boston, Massachusetts
Boley, S. J.	Montefiore Hospital and Medical Center and The Albert Einstein College of Medicine, Bronx, New York
Børresen, H. C.	Department of Clinical Chemistry and Pediatric Surgery, Rikshospitalet, Oslo, Norway
Brennan, M. F.	Department of Surgery, Harvard Medical School and Peter Bent Brigham Hospital, Boston, Massachusetts
Burri, C.	Department of Traumatology, University of Ulm, Ulm, West Germany
Chance, G. W.	The Hospital for Sick Children, Toronto, Ontario
Cohen, M. I.	Department of Pediatrics, Montefiore Hospital and Medical Center and The Albert Einstein College of Medicine, Bronx, New York

Crawford, J. D.	Department of Pediatrics, Harvard Medical School, Massachusetts General Hospital and Shriners Burns Institute, Boston, Massachusetts
Daum, F.	Montefiore Hospital and Medical Center and The Albert Einstein College of Medicine, Bronx, New York
Dolif, D.	Department of Medicine, St. George General Hospital, Hamburg, West Germany
Driscoll, J. M., Jr.	Department of Pediatrics, Columbia University College of Physicians and Surgeons, Babies' Hospital and Columbia-Presbyterian Medical Center, New York, New York
Dudrick, S. J.	Department of Surgery, The University of Texas Medical School, Houston, Texas
Fischer, J. E.	Department of Surgery, Harvard Medical School and Massachusetts General Hospital, Boston, Massachusetts
Förster, H.	Department of Biochemistry and Physiology, J. W. Goethe University, Frankfurt, West Germany
Geyer, R. P.	Department of Nutrition, Harvard School of Public Health, Boston, Massachusetts
Greene, H. L.	Metabolic Division, U. S. Army Medical Research and Nutrition Laboratory, Fitzsimmons General Hospital, Denver, Colorado
Hambidge, M.	Metabolic Division, U. S. Army Medical Research and Nutrition Laboratory, Fitzsimmons General Hospital, Denver, Colorado
Hegstedt, D. M.	Department of Nutrition, Harvard School of Public Health, Boston, Massachusetts
Heird, W. C.	Department of Pediatrics, Columbia University College of Physicians and Surgeons, Babies' Hospital and Columbia-Presbyterian Medical Center, New York, New York
Heller, L.	Department of Obstetrics and Gynecology, J. W. Goethe University, Frankfurt, West Germany

CONTRIBUTORS

Herman, Y. F.	Metabolic Division, U. S. Army Medical Research and Nutrition Laboratory, Fitzsimmons General Hospital, Denver, Colorado
Herrin, J. T.	Department of Pediatrics, Harvard Medical School and Massachusetts General Hospital, Boston, Massachusetts
Hofert, C.	Children's Hospital Borgfelde, Hamburg, West Germany
Jürgens, P.	Department of Medicine, St. George General Hospital, Hamburg, West Germany
Knutrud, O.	Department of Pediatric Surgery, Rikshospitalet, Oslo, Norway
Law, E. J.	Department of Surgery, University of Cincinnati Medical Center, Cincinnati, Ohio
Litt, S. J.	Montefiore Hospital and Medical Center and The Albert Einstein College of Medicine, Bronx, New York
MacFadyen, B. V.	Department of Surgery, The University of Texas Medical School, Houston, Texas
MacMillan, B. G.	Department of Surgery, University of Cincinnati Medical Center, Cincinnati, Ohio
McCance, R. A.	Sidney Sussex College, Cambridge, England
Mezey, E.	Department of Medicine, Baltimore City Hospital and Johns Hopkins University School of Medicine, Baltimore, Maryland
Munro, H. N.	Department of Nutrition and Food Science, Massachusetts Institute of Technology, Cambridge, Massachusetts
Pässler, H. H.	Department of Traumatology, University of Ulm, Ulm, West Germany
Panteliadis, C.	Children's Hospital Bargfelde, Hamburg, West Germany
Popp, M. P.	Department of Surgery, University of Cincinnati Medical Center, Cincinnati, Ohio

Schonberg, S. K.   Montefiore Hospital and Medical Center and The Albert Einstein College of Medicine, Bronx, New York

Warshaw, J. B.   Department of Pediatrics, Obstetrics and Gynecology, Yale University School of Medicine, New Haven, Connecticut

Widdowson, E. M.   Dunn Nutritional Laboratory, University of Cambridge and Medical Research Council, Cambridge, England

Winters, R. W.   Department of Pediatrics, Columbia University College of Physicians and Surgeons, Babies' Hospital and Columbia-Presbyterian Medical Center, New York, New York

AUTHOR INDEX

Aas, M., 89
Abaci, F., 261
Abderhalden, E., 2, 178
Abel, R. M., 220, 225, 228
Abbott, W. M., 220, 225
Abrams, M. E., 44
Acquilera, C. T., 272
Adam, D. H. D., 182
Adam, P. A. J., 61
Ahrens, R. A., 33
Alberti, K. G. M. M., 56
Albritton, W. L., 261
Alexander, J. W., 271
Alexis, S. D., 20
Alhough, I., 80
Allen, T. R., 162
Allison, S. P., 50, 65
Alpen, E. L., 273
Altman, R. P., 269
Anderson, A. O., 272
Anderson, B. R., 275
Anggard, L., 113
Antoon, A., 247-248
Aoki, T. T., 55, 65
Aprahamian, H. A., 126
Arenalo, N., 206
Ariaey-Nejad, M.R., 139
Arias, I. M., 89
Artz, C., 245
Asch, M. J., 155
Ashcraft, K. W., 159
Ashmore, J., 51
Ashworth, A., 33, 147, 149
Ashcraft, K. W., 246, 269
Atwood, R. P., 271, 272
Auerbach, V. H., 206
Augenfeldt, J., 92
Babson, S. G., 112

Baehner, R. L., 275
Baens, G. S., 35, 50
Baerti, J. M., 133
Baertl, J. M., 33
Bailer, E., 92
Baird, J. D., 38
Baker, E. M., 139
Baker, G. L., 12, 263
Balachi, M., 139
Balasse, E., 60, 61
Balestrieri, C., 183, 192
Baliga, B. S., 14, 20
Balis, M. E., 172, 183
Ballard, F. J., 63
Bansi, H. W., 179
Bansmer, G., 271
Barness, L. A., 206
Barnett, L., 246
Barnett, G. O., 225
Barta, E., 92
Bassler, K. H., 71, 77, 85, 181
Batchelor, A. A., 239
Bates, M. W., 58, 64
Baum, D., 96
Beauchamp, C. J., 269
Beck, C. H., 225
Becker, D. E., 15
Belinkoff, S., 113
Bentley, D. M., 270, 271
Berg, G., 85
Berger, J. E., 89
Bergland, R. M., 231
Bergstrom, J., 55
Bernard, R. W., 159, 269, 272, 273, 274
Bickel, H., 85
Biebuyck, J. F., 54-67
Bigger, J. W., 272

Bishop, H. C., 151, 177, 178, 181, 182
Bjordal, R., 165-177
Bjorntorp, P., 61
Blackfan, K. D., 187
Blocker, T. G., 239
Blocker, U., 239
Blystad, W., 166
Bode, H. H., iii-v, 1-3, 247-248
Boeckman, C. R., 231, 269, 270, 272, 274
Boelsche, A. N., 182
Bogen, J. E., 271
Boley, S. J., 215
Bonnichsen, R. K., 115
Booth, G. H., 132
Boreus, L. O., 116
Børresen, H. C., 163, 165-177, 178, 181, 182
Boyer, A., 172, 183, 188
Braun, P., 271
Brayton, R. G., 270
Bremer, J., 90, 92
Brennan, M. F., 269-278
Bressler, R., 92
Breuer, E., 92
Bridgers, W. F., 139
Brink, N. G., 115
Brodie, B. B., 116
Brookes, I. M., 14
Browder, A. A., 137
Brown, P. R., 14
Bryan, H., 39
Bowie, M. D., 38, 39
Buchele, H., 76
Buckley, H. R., 272
Buergel, N., 139
Bunker, J. P., 126
Burch, H. B., 61
Burk, R., 132
Burke, J. P., 271
Burri, C., 250-255, 271
Butcher, R. W., 89
Butler, A. M., 119, 121, 125, 187
Butterfield, W. J. H., 44
Cabak, V., iv
Cahill, G. F., 19, 20, 55, 65
Caldwell, M. D., 264

Calkins, L. A., 115
Calloway, D. H., 36
Canham, J. E., 139
Cardon, J., 273
Carlson, L. A., 66
Carter, E. A., 113
Cassady, G., 47
Chamberlain, M. J., 50
Chase, H. P., 188, 189
Chase, R. M., 159
Choi, S. C., 275
Clifford, A. J., 14
Cittadini, D., 183, 192
Coats, D., 2, 179, 182
Cohen, M. I., 215, 216, 262
Colcher, H., 50
Conard, V., 44
Coon, M. J., 33
Coran, A. G., 163, 176, 178, 181
Cordano, A., 33, 133
Cornblath, M., 38, 39, 45, 50
Couturier, E., 60
Craig, J. W., 55
Cravioto, J., iv
Crawford, J. D., 119-130
Crigler, J., 214
Curet, L. B., 206
Curreri, W., 245
Curry, C. R., 159, 269, 271, 272, 274
Cuthberton, D. P., 18, 64, 65
Daae, N., 89
Dalton, H. P., 271, 275
Daly, J. M., 151
Dancis, J., 172, 183, 188
Darby, F. J., 115
Darby, W. J., 132
Das, J. B., 161
Dascomb, H. E., 273
Davidson, C. S., 239
Davies, J. P., 165
Davies, P. A., 165
Davis, H., 182
Dawkins, M. J. R., 61
Decarli, L. M., 113, 115
De Castro, A. F., 206
Dehmel, K. H., 71
Dekker, E. E., 225
DeLicardie, E. R., iii

AUTHOR INDEX

Dell, R. B., 261
Dennis, D. C., 271
Denny-Brown, B. C., 271
Denton, A. E., 13
Diamant, E. J., 61
Dickson, M., 39
Dinwoodie, A. J., 55
Dobbing, J., iii
Dodd, K., 50
Dodge, P., 126
Doisey, R. J., 133
Dolif, C., 178-198
Donnell, G. M., 50
Doolan, P. D., 273
Dormandy, T. L., 50
Driscoll, J. M., Jr., 215, 216, 261, 262
Drucker, W. R., 55
Druskin, M. S., 271, 272
Dudrick, S. J., 2, 127, 151-164, 170, 178, 181, 182, 220, 225, 227, 241, 256-268, 269
Duhig, J. V., 272
Duma, R. J., 271, 275
Dweck, H. S., 47
Edwards, J. B., 80
Edwards, R. G., 80
Eichner, E. R., 139
Eisenberg, C. S., 275
Elvehjem, C. A., 13
Elwyn, D., 12, 13
Eraklis, A. J., 161, 231
Erdmann, G., 181, 182
Exton, J. H., 62, 64
Farquhar, J. W., 38, 39
Farr, L. E., 2, 178, 187
Farrigus, U. S., 14
Fauconneau, G., 12, 23
Geigin, R. D., 275
Feinstein, R. N., 115
Fekety, F. R., 271
Felber, J. P., 51
Felig, P., 19, 20
Fikrig, S. M., 275
Filer, L. J., 28
Filler, R. M., 161, 231, 275
Finberg, L., 166
Firth, J. A., 33
Fischer, J. E., 220, 225-230
Fishman, B., 13
Fitts, C., 246
Flake, K. K., 275
Fletcher, W. S., 271
Flint, J. M., 163
Folkman, J., 214
Fomon, S. J., 5, 28
Forsander, O. A., 79
Förster, H., 71-87
Foster, G. S., 220
Franckson, J. R. M., 60
Frank, F., 178
Franklin, F. A., 275
Fredrickson, D. S., 56
Freedland, R. A., 60
Freeman, L. R., 271, 273
Freeman, R., 272, 275
Fritz, I. B., 90, 95
Froesch, E. R., 85
Fuchs, A. R., 112
Fuchs, F., 112
Gamble, J. L., 119, 121, 125, 128
Farland, P. S., 56
Garrison, F. H., 3
Gaull, J. G., 263
Gentz, J., 39, 45
Geser, C. A., 71
Geyer, R. P., 2, 98-111
Ghadimi, H., 261
Giordano, C., 183, 192, 225
Giovanetti, S., 225
Gitler, C., 12
Gitlin, D., 206
Gliedman, M. L., 215
Godard, C., 49
Goldman, M. H., 270, 274
Goodman, M. J., 162
Gopalan, C., 33
Gordon, R. S., 56
Graham, G. G., 30, 33, 133
Grasso, S., 43
Grebin, B., 215, 216
Greene, H. L., 131-145
Greenstein, J. P., 184, 189, 190
Griem, W., 181
Groff, D. B., 151, 168, 178, 181, 182, 231
Guerrero, J., 113, 114

Guzman, M. G., 272
Gyorgy, P., 183, 188
Hales, C. N., 56
Hall, O. W., Jr., 113
Hallberg, D., 175
Hambidge, K. M., 131-145
Hammel, D., 206
Hammond, J., 5
Hansen, A. E., 182
Harmon, B. G., 15
Harper, A. E., 14, 190
Harper, H. A., 273
Hartmann, G., 181
Hasenclever, H. F., 270
Haslbeck., M., 71, 76
Hause, N. L., 188
Havel, R. J., 62
Hawkins, R. A., 58, 64
Hays, D. M., 155
Heesen, D., 77
Hegstedt, D. M., 34, 166, 182
Heimberg, M., 60
Heinz, F., 72
Heird, W. C., 215, 216, 256-268
Heller, L., 179, 206-213
Hems, R., 59, 60
Henderson, E. S., 270
Herman, R. H., 139, 227
Herman, Y. F., 131-135, 139
Hofert, C., 196-198
Hoggard, M. E., 182
Hollmann, S., 72
Holt, L. E., 27, 28, 31, 34, 172, 175, 183, 188
Hillman, R. S., 139
Hinton, P., 50
Hirsh, J., 5
Ho, O. L., 12
Hodgis, R. E., 139
Hoffbrand, A. V., 139
Hogan, G., 126
Holder, I., 246
Hommes, F. A., 92
Hood, J., 139
Hoos, I., 77, 80
Hopkins, M. S., 211
Hornady, G., 139
Howell, R. R., 207
Hulsmann, W. C., 88, 95
Hutchin, M. E., 273

Hutchinson, D. C., 206
Huter, J., 210
Huxtable, R. F., 155
Ikkos, D., 44
Imai, Y., 115
Irwin, M. I., 166
Isles, T. E., 39
Isselbacher, K. J., 113
Itoh, T., 64
Jensen, A. H., 15
Joh, R. V., 272
Johnson, B. C., 17, 33
Johnson, S. A. M., 272
Johnston, J. D. A., 65, 261
Jondorf, W. R., 116
Jones, S. A., 270
Joselow, M. M., 138
Jounson, H. T., 264
Jurgens, P., 178-198
Kaplan, J. H., 14
Karp, M., 112
Karpel, J. T., 138
Kass, E. H., 271
Kaye, R., 50
Keith, D., 271
Kekomaki, M. P., 56
Keller, U., 85
Kendall, F. E., 50
Kennedy, A. C., 55
Kerrigan, G. A., 121
Kiley, J., 166
King, B., 272, 275
Kipnis, D. M., 89
Kirsch, J., 72
Kjekshus, J. K., 96, 236
Klee, C. B., 64
Klenk, E. L., 258, 259
Knittle, J. L., 5
Knutrud, O., 163, 165-177, 178, 181
Kohout, M., 60
Komrower, G. M., 188, 189
Kraut, H., 188
Krebs, H. A., 54, 58, 59, 60, 64, 66
Krill, C. E., 269, 270, 272, 274
Krill, C. E., Jr., 231
Kuhlman, A. M., 61
Kumar, S., 261
Kumate, J., 206

AUTHOR INDEX

Kundsin, R. B., 270, 271, 272, 274, 275
Kuttner, R. E., 207
Lambert, G. F., 33
Lamprecht, W. L., 72
Lane, H. C., 50
Lang, K., 72, 181, 189, 190
Lark, C., 272
Lark, K. G., 272
Larsson, Y., 38
Laster, L., 188, 189
Law, E. J., 240-249
Leape, L. L., 159, 256, 269
LeDune, M. A., 39
Lee, K. L., 272
Leibowitz, S. F., 8
Lepper, M. H., 270, 271
Lerner, R. L., 39
Lerche, D., 77, 80
Levin, R. A., 133
Levin, S. L., 239
Levenson, S. M., 239
Ley, R. G., 113
Lieber, C. S., 79, 112, 113, 115
Lindseth, R. E., 66
Linton, A. L., 55
Litt, I. F., 215, 216
Lochner, A., 61
Lockwood, E. A., 92
Long, J. M., 162, 178, 181, 182, 225
Longenecker, J. B., 188
Loridan, L., 60
Lorincz, A. B., 207
Louria, D. B., 138, 270
Lowry, O. H., 61
Lowry, S. R., 61
Lucas, R. N., 269
Luft, R., 44
Luke, R. G., 55
Lund, C. C., 239
Lund, P., 66, 67
Lupin, A. M., 273
Lusk, G., 82
Luttrell, C. N., 166
MacFadyen, D. H., 151-164, 178, 187, 261
MacLachlan, F., 187
MacMillan, B. G., 240-249
Madison, L. L., 60, 61

Maenpaa, P. A., 56
Maggiore, Q., 225
Mahadevan, S., 89
Maickel, R. P., 116
Majanvik, R., iv
Malvey, P., 273
Margen, S., 36
Maroquiz, N. R., 90
Martin, J. M., 47
Matula, G., 275
Matzkies, F., 85
Mayes, P. A., 60
McArthur, J. W., 120
McCance, R. A., 4-9, 123, 146, 146, 148
McClain, L. D., 139
McClelland, R. N., 231
McGinn, M., 273
McNair, R. D., 18
Meade, R. J., 15
Mead, M., 272
Meakins, J. L., 271
Mebrane, D., 60, 61
Mehnert, H., 71, 76, 80, 181
Melichar, C. S., 181
Mellanby, J., 59
Messina, A., 43
Metta, U. C., 17, 33
Meyer, E., 77, 80, 81
Mezey, E., 112-118
Micheals, L., 272
Michel, M. C., 12, 23
Mickelsen, O., 5
Miller, M., 55
Miller, S. A., 18
Milner, R. D. G., 49
Mishkin, S., 89
Mitchell, J. R., 15
Mjøs, O. D., 96
Moore, F. D., 270, 274
Morales, C., 206
Moran, J. M., 271, 272
Morrow, G., 206
Mulligan, P. B., 38, 39, 50
Munro, H. M., 11-26
Muecke, W., 20
Majarian, J. A., 273
Nakagawa, I., 30, 31, 32
Nathan, D. G., 275
Nelson, J. H., 272

Neyzi, O., 127
Newnes, W., 115
Newsholme, E. A., 56
Nicholson, J. F., 262
Nicosia, A., 119, 123
Nigran, G., 112
Norden, L. W., 271
Noriega, L., 206
Norton, H. W., 15
Norton, P. M., 175, 188
Nowinski, W. W., 239
Nyberg, A., 113
O'Brien, John, 44, 47
Ockner, R. K., 89
O'Connell, R. C., 270, 271, 272, 274, 275
Odievre, M., 50
Oestemer, G., 15
Oliver, I. T., 63, 64
Olney, J. W., 12
Omura, T., 115
O'Neill, R., 206, 207
Ooms, H. A., 60, 61
Orme-Johnson, W. H., 113
Orrenius, S., 116
Othersew, H. B., 264
Owens, F. N., 14
Owens, J. E., 55
Page, M. A., 58, 60, 64
Panteliadis, C., 178-198
Pappova, E., 92
Park, B. H., 275
Parmley, T. H., 207
Passler, H. H., 250-255
Patek, A. J., 50
Paterson, P. Y., 275
Pawlak, M., 15
Pearse, W. H., 207
Pearson, E., 245
Peden, V. H., 138, 215
Persson, B., 38, 39, 45
Peterson, C. G., 271
Phansalkar, S. V., 175, 188
Philippart, A., 214
Pikkarainen, P. H., 113, 114
Pion, M., 15
Pipat, C., 112
Pitot, H. C., 14
Plentl, A. A., 206
Poblete, V. F., Jr.
Popp, M. P., 240-249

Porte, D., 39
Pozefsky, T., 19, 20
Pratt, E. L., 183, 188
Preisler, H. D., 270
Prellwitz, W., 181
Prenton, M. A., 206
Prince Evans, D. A., 115
Probst, J. H., 124, 125
Prochazka, J. V., 269
Prose, P. H., 183, 188
Prowse, K., 50
Pruitt, B., 140, 245
Quastel, J. H., 64
Quie, P. G., 159, 269, 271, 272, 274
Raima, N. C. R., 113, 114, 263
Raivo, K. O., 56
Ramo Rao, P. B., 17
Randle, P. J., 56
Randolph, J. G., 269
Rane, A., 116
Rathi, M., 261
Rea, W. J., 231
Reardon, H. S., 181
Reca, M. E., 272
Reitano, G., 43
Renold, A. E., 51
Rhoads, J. E., 151, 178, 181, 182, 220, 269
Richter, A. R., 92
Rickham, P. P., 181
Risk, A., 112
Riumallo, J. A., 14
Rivlin, R. S., 139
Robins, A., 188, 189
Roitman, E., 172, 175, 183, 188
Rose, W. C., 33, 182, 225
Rosenweig, N. S., 139
Rosol, J., 271, 272, 275
Rosso, P., iii
Rostad, R., 166
Rousseau, C., 273
Rowe, M. I., 271, 272
Ruberg, R. L., 162, 259
Rubin, E., 112
Rubin, V. G., 161
Ruebner, B., 272
Russell, A. D., 272
Rustishauser, I. H. E., 147, 148
Ryan, J., 220

AUTHOR INDEX

Sachs, F. L., 271
Samols, E., 50
Samonds, K. W., 34
Sanstead, H. H., 132
Saporito, N., 43
Sato, R., 115
Sauberlich, H. E., 139
Saver, F., 89
Scammon, R. E., 115
Schalch, D. S., 89
Schittenhelm, A., 178
Schmidt, G. W., 181, 182, 184
Schoen, E., 119, 123
Schonberg, S. K., 215, 216
Scholtz, R., 113
Schreier, K., 184
Schubert, O., 175
Schuberth, J., 113
Schullinger, J. N., 215, 216, 261, 626
Schumer, W., 80
Schwartz, R., 38, 39, 50
Scott, H. N., 15
Seabury, J. H., 273
Senior, B., 60
Seppala, M., 113, 114
Shackelford, P. G., 275
Sharpe, L. G., 12
Shils, M. E., 133
Shipp, J. C., 88
Shohl, A. T., 2, 178, 187
Shojania, A. M., 139
Siegal, P. D., 47, 271, 272
Silen, W., 273
Sjoqvist, F., 116
Skelton, M. A., 215
Skerjance, J., 61
Slater, J. E., 166
Smith, E. B., 17
Smithwick, E. M., 275
Smits, H., 271, 273
Snyderman, S. E., 27, 28, 31, 172, 173, 175, 183, 188, 206, 209
Sobel, E. H., 120
Soeldner, J. S., 19, 20, 66
Sokol, J. K., 112
Sokoloff, L., 64
Sornson, H., 207

Soroff, H., 245
Sotos, J., 126
Sowers, J. E., 15
Spiglano, I., 215, 216
Srikantia, S. G., 33
Spitzer, K., 245
Stahl, W. H., 159
Stahl, W. M., 269, 272, 273, 274
Stegink, L. D., 12, 263
Steiger, E., 151, 178, 181, 182, 225
Steinberg, D., 89
Stevenson, R. E., 207
Stieg, H., 184
Stifel, F. B., 139, 227
Stockland, W. L., 15
Streeton, D. M. P., 133
Sturman, J. A., 263
Sugerman, H., 259
Sunshine, P., 261
Sutherland, A. B., 239
Suzuki, T., 30, 31, 32
Swenerton, H., 133
Takahashi, T., 30, 31, 32
Talbot, N. B., 120, 121, 122, 126, 128
Tamminen, V., 113, 114
Tayler, F. H., 239
Terry, M. T., 92
Tesluk, H., 271
Thoburn, R., 271
Thomas, G. H., 207
Thoren, L., 55
Thurman, R. G., 113
Tillotson, J. A., 139
Tilstone, W. J., 65
Toala, P., 271
Tobin, J., 19, 20
Tobon, F., 114
Toussaint, W., 181
Travis, S. F., 259
Unbehaun, V., 181
Underwood, E. J., 132
Unger, R. H., 61
Usatequi-Gomez, M., 211
Van Buren, C. T., 261
Van Duyne, C. M., 62
Van Uden, N., 272
Vars, H. M., 151, 269

Vaughan, M., 89
Vidal-Leiria, M., 272
Volpe, J. J., 188, 189
Von Dippe, P., 61
von Euler, U., 38
Wagner, G., 113, 114
Wagner, L., 113, 114
Wah-Juntze, 214
Wahren, J., 19
Walker, F. A., 261
Wallace, P. G., 60
Wallace, W. M., 183, 188
Waltman, R., 112
Walton, A., 5
Warner, J. F., 271, 275
Warrner, R., 39, 45
Warshaw, J. B., iii-v, 88-97
Watson, P. R., 272
Webb, C., 256
Wei, P., 39
Weinstein, I., 60
Whichelow, M. J., 44
White, W. A., 231
Widdowson, E. M., 5, 7, 8, 123, 146-150
Wiese, R. F., 182
Wilkinson, A. W., 209
Williams, G. S., 258, 259
Williamson, D. H., 58, 59, 60, 64
Wilmore, D. W., 127, 151, 155, 168, 178, 181, 182, 227, 245, 269
Wilson, J. E., 33
Winick, M., iii, iv
Winitz, M., 184, 189, 190
Winters, R. W., 151-164, 215, 256-268
Wittles, B., 92
Winslow, P. R., 215
Womack., M., 33
Woods, H. F., 56
Woodward, H., Jr., 55
Wretlind, A., 175
Wurtman, R. J., 13
Wybregt, S. H., 38, 50
Wyrick, W. J., 231
Yaffe, S. H., 116
Yardley, J. H., 272

Yeung, D., 64
Young, M., 206
Young, V. R., 15, 16, 20
Zahnd, G. R., 49, 51
Ziege, M., 77, 80, 81
Ziegler, D. M., 113
Zimmerman, R. A., 15
Zimmermann-Telchow, H., 188
Zinner, S. H., 271
Zlatos, L., 92

SUBJECT INDEX

Acetaldehyde
  formation in fetal liver, 114
  in ethanol metabolism, 113
Acetate
  concentration in infusates, 180
  in treatment of hypochloremic acidosis, 261
Acetoacetate, 58-61
Acetoacetyl-CoA, 58-59
Acetyl CoA synthetase
  developmental increase in, 92
  in fatty acid metabolism, 90
Acid-Base
  disorders, 257, 260-261, 265-266
  during intravenous alimentation, 203, 205, 259, 265
Acidosis
  correlation with potassium concentration, 258
  from fructose infusion, 50
  hyperchloremic, 257, 260-261
  in Maple Sugar Urine disease, 237
  lactic, after galactose, 51
  relation to nitrogen balance, 203
Actin, 3-methylhistidine in, 20
Acylcarnitine
  esters of fat or acids, 91
  in fatty acid oxidation, 90
  transferases, 91, 93
Adenyl Cyclase, role in lipase activity, 89
Adipose tissue
  as fatty acid source, 60, 89
  effect of hypoxia on lipolysis, 96

Adipose tissue - continued
  effect of insulin, postoperatively, 66
Age, dependence of K_G on, 47
Alanine
  concentration in infusates, 182
  release from muscle, 19-20
  role in gluconeogenesis, 63, 65
  serum levels after amino acid infusion, 174, 194
  serum levels in newborns, 207-209
Albumin
  in serum, during starvation and refeeding, 30
  role in fatty acid transport, 89
Alcohol
  dehydrogenase, 113-116
  glucostatic theory, 6
  in parenteral nutrition, 181
  metabolism during development, 112-118
  oxidation, 112-118
  relation to hunger, 8
Alloxan, in experimental diabetes, 77
Amigen^R
  infusion of, in adults, 263
  inorganic phosphate in, 260
Amino acids
  catabolic enzymes in fetal livers, 61
  commercial preparations, 134, 172-176
  during glucose loading, 51
  effect on renal phosphate reabsorption, 259
  in abnormal liver function, 216

Amino acids - continued
  in Maple Sugar Urine Disease, 237
  infusion method for, 162, 168, 178, 198, 225-229, (in burned children), 153, 240-248
  insulin responses to, in prematures, 43
  intrauterine feeding of, 206-213
  levels during parenteral feeding, 154, 166, 232
  metabolic effects of, 178-194, 260-261, 263-264
  requirements in childhood, 27-37
  supply in renal failure, 225-226
  tolerance of, 172, 182-183
  utilization for energy, 11-26, 54, 64-65
  (also, see Specific Amino acids)
Aminofusin^R, 168, 179, 234-235
  use in burned children, 247-248
Aminograms, 175, 263-264
Aminosol^R
  liver damage, 262
  trace elements in, 156
Ammonia
  metabolism, disorders in, 257, 261-262
  serum levels during parenteral nutrition, 154-164, 206-213, 231-238, 256-265
  serum levels in newborns, 207
Amniotic fluid
  aspiration of, 180
  nutrition of fetus via injection into, 206-213
Amphotericin
  in treatment of Candida sepsis, 246, 270, 274
Anemia
  dilutional, 120
  in copper and iron deficiencies, 132
Anesthesia, fluid administration in, 126-126

Anorexia
  and zinc deficiency, 132
  in burned children, 239
  role of hypothalamus in, 8
Antibiotics
  for systemic bacteremia, 270
  local application, 156, 158, 159, 250, 254
Anticoagulation, 254
Antidiuretic hormone, 124, 232, 236
Arginine
  concentration of, and adequacy of intake, 15, 182, 190
  for treacment of hyperammonemia, 262
  levels after infusion, 174
  levels in newborns and prevention of acidosis, 207-210, 261
Ascorbic acid
  concentration in infusates, 240
  deficiency, symptoms of, 139
  requirements during infancy, 141
  used in "Renal Failure IV Diet", 226
  (also, see Vitamins)
Aseptic technique, 154-156, 254
Aspargine
  separation from glutamine, 192
  serum levels after infusion, 174
Aspartic acid
  concentrations in infusates, 183, 192
  serum levels after infusion, 174
  serum levels in newborns, 207-210
  transamination of, in intestinal mucosa, 12
Aspergillosis, sepsis from, 270
ATP (adenosine-tri-phosphate)
  in fructose phosphorylation, 56
  in liver after halothane, 66
Atrium, relation of catheter tip to, 251
Azotemia, during intravenous alimentation, 204, 257, 260
Basal Metabolic Rate (BMR), as parameter for amino acid needs, 30

SUBJECT INDEX

Behavior, as related to energy
  intake, 8-9
Beta-Adrenergic Stimulation
  role of in lipase activity, 89
Bile
  as endogenous protein source,
    12
  stasis during parenteral nutrition, 262
Bilirubin
  binding to proteins, 89
  serum levels, 215, 216
  serum levels after infusion
    programs, 77, 80, 81, 85,
    262
Biotin, daily recommendation
  for, 141
Blood
  cultures, 241, 270, 274
  post natal changes, 62, 64
  studies during parenteral nutrition, 154-156
  withdrawal and administration
    of, 159
Blood Urea Nitrogen (BUN)
  during parenteral nutrition,
    41, 154, 204, 205, 227, 230,
    241, 265
Bone
  abnormalities in copper deficiency, 132
  growth in neonates, 166
Bowel
  adaptation of wall thickness,
    163
  disease and parenteral alimentation, 160-162, 214-224,
    235, 238
  surgery, post-op alimentation,
    233
Brain
  glucose requirements, 54, 85
  incomplete development, 263,
    264
  size and malnutrition, 98
  utilization of ketone bodies,
    60, 64
Brain stem, role in nutrition, 6
Bromo-palmitate, as fatty acid
  analogue, 89

Burns
  acute renal failure in, 226
  glucose/insulin ratio in, 65
  parenteral nutrition in, 239-
    249, 235
  septic risk, 269
Butyrine, serum levels after infusion, 174
Calcium
  abnormal metabolism of, 257-260,
    265-266
  absorption, 153
  concentration in human milk, 167
  concentration in infusates, 152,
    156, 173, 180, 181, 200, 216
  deposition in bone, 167
  plasma levels after alimentation, 154, 204, 205, 241
  retention, 171
  role of in absorption of trace
    elements, 132
Caloric intake
  and hyper glycemia, 65
  and weight gain, 147, 200
Calories
  concentration in infusates,
    112, 127, 151-152, 167, 199,
    202, 227
  fatty acids as source, 88
  homeostasis, 56-60, 67
  relationship to energy expenditure, 6
  requirements for, 33, 34, 129,
    148, 149, 178-182, 231-232,
    234, 237, 239, 241-246
Candida catheter sepsis, 244,
  270-275
Canula, split, 252
Carbohydrate
  as a direct fuel, 55-60
  concentration in infusates, 149,
    152-153, 167-168, 173, 181,
    248
  content in human milk, 167
  effect on plasma amino acid
    levels, 18
  intravenous tolerance of, 38-53,
    62
  metabolism following anesthesia,
    65-67

Carbohydrate - continued
  requirements, 231
Carbon dioxide, partial pressure as related to nitrogen balance, 203
Carbon tetrachloride, damage to liver explants, 262
Carboxykinase, 63
Carboxylase, 63
Cardiac failure
  altered water and salt tolerance in, 233
  in hyperosmolar coma, 258
Carnitine
  role in fatty acid oxidation, 91-93
Catalase, determinations of in fetal liver, 114-116
Catecholamines, in inhibition of insulin release, 50
Catheter
  cave, 250
  for parenteral alimentation, 155-163, 180, 234-238, 247, 250
Cationic amino acids, 261
Central Nervous System
  disease, altered H_2O and salt tolerance in, 233
  incomplete development of, 263-264
Cholestasis, in patients on alimentation, 216
Chloride, concentration in infusates, 152, 180, 181
Chlorine, abnormal metabolism, 257, 260-261, 265
Choline, dietary recommendation, 141
Chromium, as required trace element, 131-138
Chromatography, column, for amino acids, 188, 192, 207, 209
Chylomicron, 89, 172
Cirrhosis, parenteral nutrition in, 215
Citric Acid Cycle enzymes, increase after birth, 91

Citrulline, serum levels after infusion program, 174
Clotting, as related to vitamin K, 140
Cobalt, recommendations for, 131, 182
Coenzyme A esters, role of in fatty acid metabolism, 89-91
Colitis, parenteral nutrition in, 220
Convulsions, associated with hyperammonemia, 262
Copper
  deficiencies in, 264
  requirements, 131-138, 152
Corticosteroids
  use of for ileocolitis, 220
  influence of on fuel utilization, 55, 60
Creatinine
  concentration in amniotic fluid, 209
  values in urine, 180
Cultures
  of IV tubing and infusates, 159
  blood, 241, 270-274
Cyclic AMP
  as related to lipase activity, 89
  influence of on fuel utilization, 55
Cystathionase, role of in conversion of methionine, 263
Cystine
  concentration in parenteral solutions, 183
  requirements of, 183
  serum levels after infusion, 174-175, 189-190
  serum levels in newborns, 207-210
Cytochrome Oxidase
  in calf intestine, 95
  in heart and liver after birth, 91-92
Cytochrome (P-450), determinations of in fetal liver, 114-116
Dehydration
  from metabolic disease, 236

Dehydration - continued
  from osmotic diuresis, 161, 256
  from unreplaced fluid loss, 129
Demerol, cause of antidiuresis, 125
Dextrose, in parenteral solutions, 39, 43, 47, 151, 153-155, 160, 162, 226, 234, 240
Diabetes
  disturbed glucose untilization in, 85
  fluid homeostasis, 120, 123
  in chromium deficiency, 133
  use of central venous catheters in, 250
Diarrhea, from duodenal tube feeding, 239
Dicarbonic acid, 111, 192
Diet
  as related to energy expenditure, 5
  average protein intake, 12
  water tolerance and requirements, 128
Digestive tract
  functions of, during illness, 129
  role of, in response to a protein meal, 11-13
Dihydroxyacetone phosphate, role of in sorbitol metabolism, 71-72
2-3-Diphosphoglycerate (2-3 DPG), levels of in hypophotemia, 259
Distal Convuluted Tubule
  reabsorption of phosphate, 259
  role in urine concentration, 124
Diuresis
  depression of by surgery, anesthesia, and opiates, 125
  osmotic, in hyperglycemia, 128, 161, 204, 237, 256-257
  therapeutically induced, 236
DNA (deoxyribonucleic acid), suppressed thymidine incorporation during zinc deficiency, 133

Dressings, sterile, occlusive, to IV sites, 156, 158, 159
Drug Therapy, 232-233
Duodenum
  mucosal thickness and villus height, 163
  stenosis and parenteral nutrition, 169
  tube feeding in burned children, 239
Edema
  as related to water intake, 203
  pulmonary, in hyperosmolar coma, 258
  nitrogen retention in, 186-187
Eiconsatrienoic Acid, increase of, with fatty acid deficiency, 264
Electrocardiograms, abnormal, with Vitamin B_6 deficiency, 139
Electroencephalograms, 139
Electrolytes
  in alimentation solutions, 135, 151, 168, 172-175, 200
  metabolism, disorders of, 256-259, 265
  requirements and tolerance, 119-129, 231-232, 234-238
  serum values during infusion, 156, 180, 241
Electrophoresis
  protein, of hyperalimented infants, 265
  starch gel, of adult, newborn, and fetal livers, 114
Elements, trace, nutritional importance of, 131-145, 181
Embolism, from central venous lines, 159, 231, 248, 251-252
Encephalitis, reduced water and salt tolerance in, 233
Endocarditis, from infected catheter tip, 135
Endoplasmic reticulum, 89
Energy, requirement, 88
Enteritis, regional, 220
Enterocolitis, necrotizing, 199
Enzymes (see also specific enzymes)
  activity, variations during development, 206

Enzymes - continued
  adaptation, 167, 188-189
  digestive, 146
  in fatty acid metabolism, 89-96
  required for degradation, 54
Epinephrine, influence on fuel
   utilization, 54, 60
Erythrocyte
  dependence on glucose, 54
  inorganic phosphate compounds
   of, 259
Ethanol
  administration, effect on
   metabolism, 79
  metabolism during development,
   112-118
  (also see alcohol)
Failure to Thrive, 233
Fat
  availability of emulsions, 163
  disposition after alimentation,
   203
  increased tolerance, after
   trauma, 66-67
  infusion, protective effect
   against thrombophlebitis,
   239, 247-248
  intake and dietary preferences,
   5
  interrelationships with carbohydrate metabolism, 56-60, 62
  parenteral administration of,
   67, 149, 151-153, 163, 165-
   177, 181-182
  tissue accumulation, 172
Fatty acids
  as energy source, 54-55
  deficiency in, 264
  fuel preference by certain
   tissues, 88
  levels, 47-49, 77-78
  metabolism, role of anesthesia
   on, 66-67
  oxidation, 88-97
  role in ketogenesis, 60-64
Fatty infiltration, after parenteral nutrition, 262
Feces
  protein content, 12
  nitrogen losses in, 13

Fetus
  energy intake, regulation of,
   4-5
  protein requirements, 6
  intrauterine amino acid feeding
   of, 206-213
Fever
  as indicator of sepsis, 159,
   254-255
  energy expenditure during, 5
  salt and water tolerance during,
   232-233
  therapeutic control of, 232-233
  water requirements during, 123
Fibrosis, in liver, after parenteral nutrition, 262
Filter, micron membrane in catheters, 156, 158
Fistulae
  parenteral nutrition in patients
   with, 223, 235
  tracheo-esophageal, 233
Fluid, requirements and tolerance,
   119-129, 231-233, 234
Fluoroscopy, chest, for catheter
   tip placement, 156-157
Folate
  activity of in cell replication,
   131
  deficiency, 193
  dietary recommendation, 141
  increased requirement during
   brain development, 139
Folic acid, in parenteral solution, 153-154, 156, 264
Food
  intake, control over, 5
  supply at birth, 4
Food and Agriculture Organization,
   27, 29, 32
Food and Drug Administration, 152
Fre Amine[R]
  inorganic phosphate in, 260
  metabolic disorders from, 260-
   261, 266
  trace elements in, 134-135
Fructose
  parenteral use of, 50-51, 55-56,
   152-153, 247-248
  role in glyconeogenesis, 63

SUBJECT INDEX

Fructose - continued
role in xylitol metabolism, 72
utilization of, 71-86
Fructosuria, 50
Fuel, supply and regulation, 54-67
Fungus, IV line sepsis from, 168, 270-275
Galactose
administration, 79
intravenous tolerance of, 50-51
Galactosemia, 214
Gastrointestinal tract
absorption of vitamins from, 142-143
delayed emptying of, 239
juices, as endogenous protein source, 12
Gavage Feeding, 236
Gestational Age
as related to energy intake, 4
as related to glucose tolerance, 44-48
Glomerulus, 259
Glomerulonephritis, parenteral nutrition in, 215, 226
Glucagon, influence on fuel utilization by tissues, 55, 60
Gluconeogenesis
in neonates, 61, 66
inhibition of after anesthesia, 66-67
reduction of after carbohydrate infusion, 78, 82-83
role of in metabolism of fructose, xylitol and sorbital, 71, 75
source of carbon for, 19
Glucose
blood levels after infusion of solutions, 38, 40, 43, 61-62, 64, 73-85, 154, 161, 265
effect on plasma free amino acids, 18
inability to handle, 204, 205
intravenous tolerance in infancy, 38-50
metabolism, disorders in, 256-259, 264-266

Glucose - continued
oxidation of in contrast to fatty acids, 56-60, 92, 95
parenteral use of, 38-50, 112, 127, 134, 179, 200, 225, 228-229, 247-248
role in nitrogen catabolism, 64-67
tolerance and intolerance, 39, 43, 65-67, 132-134, 153
transplacentally derived, 91
urinary losses, 231, 246
utilization, 71, 54
Glucostatic theory, 6
Glucuronidation, of bilirubin, after ethanol infusion, 112
Glutomate-Malate oxidation, in neonates, 92
Glutamic Acid
activity of in rat liver, 14
as related to Candida growth, 272
concentration in infusates, 183, 192
serum levels after infusion, 174
serum levels in newborns, 208-210
transamination of, in intestinal mucosa, 12
Glutamine
determination, 192
role in gluconeogenesis, 65
serum levels after infusion, 174
serum levels in newborns, 207
Glyceraldehyde, role of utilization of sorbitol, 71-72
Glyceric acid kinase, from phosphorylated glyceraldehyde, 72
Glycerol
blood levels after ketone infusion, 61
role in gluconeogenesis, 63-64
Glycine
concentrations in parenteral solutions, 182, 192, 194
role of in GPT activity, 216
serum levels after infusion, 174, 263
serum levels in newborns, 207-210

Glycogen
  from metabolized xylitol, sorbitol and fructose, 71-86
  release, 54-55
  storage disease, nutrition of children with, 215
Glycogenolysis, liver, contribution of glucose, 62, 85
Glycogenosis, 214
Glycolytic enzymes, hepatic, 139
Glycolytic pathway, 61, 63, 72
Glycopeptide, relationship to anorexia, 6
Glycosuria
  as related to onset of infection, 50
  associated with hyperglycemia, 161, 204, 256
  examination and extent of, 41-42
GOT, serum elevation of, 216
GPT, abnormal elevation in, 216
Growth
  after amino acid diets, 28, 31, 34-36
  as related to hunger, 8
  during total parenteral nutrition, 151, 163
  elements needed for, 231
  in patients with renal failure, 226
  protein requirements, 31-33
  retardation, in zinc deficiency, 132-133
  spurts, in undernourished children, 149
Halothane, effect on gluconeogenesis, 66-67
Hair, trace elements in, 137-138
Head, circumference, 47, 154, 172
Heart
  ability to utilize ketone bodies, 60
  effect of fat emulsions on function, 96
  failure, with fluid overloading, 241
  perforation, by catheter tip, 155, 251, 252

Heart - continued
  preference for fatty acids, 88-92, 54
Hematopoiesis, function of Vitamin E, 140
Hemoglobin, oxygen dissociation curve of, 259
Hemodilution, 120
Heparin
  effect of infusions on triglyceride levels, 175
  in fatty acid release, 89
Hepatobiliary Disease, role of parenteral nutrition in, 214-220
Hepatocyte, 89
Hepatomegaly
  associated transaminase levels, 262
  resulting from parenteral infusion, 50, 216
Hepatotoxicity, lack of after carbohydrate infusion, 81
Hexosemonophosphate Shunt, in neonatal livers, 61
Histidine
  concentration in infusates, 182, 190, 226-227
  methylation of, in skeletal muscle, 20
  serum levels after infusion programs, 174
  serum levels in newborns, 207-210
History, of parenteral nutrition, 1-3
HMG CoA
  lyase reaction, 58-60
  synthase reaction, 58-60
Homogenates
  heart, 92
  intestinal mucosal, 95
  liver, 113
Hemostasis, 122, 127
Humidification
  therapeutic, 233
  water loss from, 236
Hyaline Membrane Disease, 180

SUBJECT INDEX

Hydrogen
  generated by metabolism of cationic amino acids, 261
  peroxide, role in oxidation of ethanol, 113
Hydrolysates
  as related bo abnormal liver function, 216
  brain damage from, 12
  metabolic complications of, 260-261
  use of in parenteral solutions, 39, 112, 127, 134, 138, 151, 153-154, 155, 178, 187, 200, 216, 240, 247-248, 272
Hydropenia, 124
Hydroxybutyrate dehydrogenase
  activity in infant brain, 64
  reaction, 58-61
Hydroxymethlyglutaryl-CoA, 58
Hydroxyproline, urinary excretion of, 192
Hydroxylase, aniline, in fetal livers, 114-116
Hyperammonemia, during parenteral nutrition, 257, 261-262
Hyperbilirubinemia, 112
Hypercalcemia, 257, 259-260
Hyperchloremia, 257, 260-261
Hyperglycemia
  during parenteral alimentation, 41, 43, 51, 65, 161, 200, 228, 256-257
  lack, following infusion of xylitol, sorbitol and fructose, 85-86
  seizures, 128
Hyperglycinemia, 263
Hyperketonemia, 64
Hypermagnesemia, 257, 259-260
Hypernatremia, 256
Hyperosmolar Coma, during intravenous alimentation, 231, 236, 257-258
Hyperphagia, hypothalamic control of, 7
Hyperphosphatemia, 257, 259-260
Hypersensitivity reaction, 231
Hypocalcemia, 257, 259-260

Hypogeusia, in zinc deficiency, 132
Hypoglycemia
  neonatal, 38, 39, 62-63
  prevention of, 41
  "rebound" after withdrawal of infusate, 50, 160
Hypomagnesemia, 257, 259-260
Hyponatremia
  during antidiuresis, 125
  from parenteral infusion, 256-259
Hypophosphatemia, 257, 259-260
Hypoproteinemia, 258
Hypothalamus
  in regulation of energy balance, 608
  osmoreceptors, 127
Hypothyroid, goiter, from iodine deficiency, 132
Hypoxemia, limiting fatty acid oxidation, 95-96, 237
Iliocolitis, nutrition of patients with, 218, 220
Ileitis, alimentation of infants with, 233
Immunosupression, in relation to IV line sepsis, 270
Inanition, 166-167
Infection
  associated with catheters, 159, 254-255, 269-278
  as related to weight gain, 148
  increased caloric demands, 239
  monitoring of patients with, 265
Insulin
  cofactor action of trivalent chromium, 133
  effect on amino acid uptake, 18-20
  effect on metabolism of xylitol, sorbitol and fructose, 73
  influence of on fuel utilization, 55, 61
  need during carbohydrate infusions, 71, 85, 246
  plasma levels during parenteral nutrition, 161
  plasma levels during stress, 65

Lipid, infusion of, 163, 181-182
Lipidosis, nutritional deficiency in, 214
Lipofundin^R, 179, 181
Lipogenesis, in neonatal liver, 61
Lipolysis
 during lipid infusion, 175
 inhibition of, 61
Lipoprotein, role in lipolysis, 60, 89
Lipostatic theory, 7
Liver
 effect of halothane anesthesia on, 66-67
 effect of parenteral nutrition on, 214-224, 262-263, 265-266
 failure and fluid tolerance, 233
 function studies, 265
 metabolism of ammonia in, 261
 mitochondria in, 91
 pathways of ketogenesis, 58-61, 64
 release of glucose from, 62, 65, 258
 role in alcohol metabolism, 54, 89, 96
 role in amino acid metabolism, 12-15, 19
 role in fatty acid metabolism, 54, 89, 96
 role in fructose phosphorylation, 56
 role in metabolism of xylitol, sorbitol and fructose, 71-81
 vitamin storage in, 139, 143
Low birth weight infants
 abnormal blood sugars in, 40
 alimentation of, 199-205
Lymphatics, 89
Lyseric acid, phosphorylation of, 72
Lysine
 concentrations in infusates, 183, 190, 226, 261
 metabolism in muscle, 19
 range of estimated requirements, 28-31

Lysine - continued
 serum levels, 174, 188, 207-210
 utilization of, 15
Magnesium
 concentration in infusates, 149, 152, 158, 167, 180, 181, 200, 216
 metabolism, disorders in, 257-260, 265-266
Malabsorption syndrome, alimentation in, 233, 235
Malate, in parenteral solution, 180
Malformations, congenital, 133
Malignancy, alimentation of infants with, 233
Malnutrition
 post burn, 239-248
 protein-caloric, 133, 165
Manganese, requirements of, 131-138, 152
Maple Sugar Urine Disease, 233, 237
Marasmus, 8, 49, 146-148
Meningitis, antidiuresis in, 233
Mesenteric Artery Syndrome, 242
Metabolic studies, during parenteral nutrition, 163, 165-177, 231, 256-268
Metabolism
 complications of, from parenteral nutrition, 256-268
 effects of amino acids on, 178
 hypermetabolic states, 239
 in glycogen storage disease, 215
Methionine
 in parenteral solutions, 174, 183, 226
 influence of fibrin hydrolysates, 272
 requirement of, 28, 31
 serum levels after infusion, 188-190, 263
 serum levels in newborns, 207-210
Methylhistidine, as non-reutilizable amino acid, 85-86
Michaelis-Menten Constant, increase in fetal liver, 114

Insulin - continued
  relation to hunger, 8
  secretion, 6, 39-43, 47-51, 258-259
Intestine
  disorders of, and nutrition, 165, 178
  motility during hunger, 8
  need for glucose, 85
  relation to fatty acid transport, 60, 89
Intralipid^R
  preparation and use of, 168, 175-176, 234-237, 247-248, 264
  serum triglyceride levels during infusion, 175
Intrauterine feeding, of amino acids, 206-213
Iodine, requirement, 131-134, 152
Iron, requirement, 131-138, 152, 156
Isoleucine
  estimated requirements, 28, 31
  in Maple Sugar Urine Disease, 237
  in parenteral solutions, 183, 190, 226
  serum levels after infusion, 174
  serum levels in newborns, 207-210
Keratogenesis, in zinc deficiency, 132
Ketoacidosis
  water requirements during, 123
  utilization of fructose
Ketogenesis
  in neonatal development, 61-64
  pathways of in liver, 58-60
  regulatory factors in, 60
Ketones
  administration of, 60-61, 67
  formation of, 58-61, 63
  utilization and metabolism of, 54, 64

Ketosis, prevention of, 231
Kidney
  dependence on free fatty acids, 54
  elimination of ethanol from, 113
  excretion of phosphate from, 259
  excretion of sorbitol by, 83
  function during hyperalimentation, 154
  loss of vitamins through, 143
  renal failure IV diet, 225-228
  role in acidosis and hyperkalemia, 258-259
  role in water and solute regulation, 121-122, 166
  utilization of ketones, 60
Krebs-Henseleit Cycle, enzymes of, 63
Kussmaul Respiration, water requirements during, 123
Labor, prevention by ethanol, 112
Lactalbumin, 34
Lactate
  as source of energy, 54, 63
  from sorbitol and xylitol, 74
  in gluconeogenesis, 63-64
  in treatment of hyperchloremic acidosis, 260-261
  levels after carbohydrate infusion, 56, 77-79, 85
  pyruvate ratio, 71, 75, 114
Lactogen, in placental insufficiency, 211
Laminar Flow Hood, for preparation of solutions, 154, 275
Leucine
  concentration in infusates, 183, 190, 226
  estimated requirements, 28, 31
  in Maple Sugar Urine Disease, 237
  role in SGPT activity, 216
  serum levels after infusions, 174
  serum levels in newborns, 207-210
Leukopenia, in copper dificiency, 132
Lineoleic acid
  deficiency, 176
  plasma levels, 264
Lipase, 89

Microsomes, ethanol oxidizing activity of, 114
Milk
 infusion of, 149, 163, 167, 169
 vitamin E content in, 140
Mineral
 concentration in infusates, 131-145, 151, 162, 264
 metabolism, disorders in, 257-260
Mitochondria
 in fatty acid metabolism, 90-92
 increase in, post birth, 92, 95
Molybdenum, 131
Monitoring, during parenteral nutrition, 265
Monoglycerides, 89
Morphine, causing antidiuresis, 125
Mortality rate, in parenteral nutrition, 160
Muscle
 amino acid metabolism in, 18-21
 dependence on fatty acids, 54
 insulin effect on, 65, 67
 resynthesization of alanine in, 65
Myocardium, function during hypoxia, 96
Myosin, 3-methylhistidine in, 20
NAD, in metabolism of poly-alcohols, 74, 79
NADPH
 role in ethanol oxidation, 113-115
 role in ketogenesis, 58
Nasogastric tube feeding, 234
Nausea, 216
Necrosis, acute tubular, 227
Neo Aminosol^R, acid-base disorders from, 260-261, 266
Nephrotic Syndrome
 parenteral nutrition in, 226
 water infusion in, 126-127
Neurohypophysis, ethanol effect on, 112
Newborns
 controlled alimentation of, 178-191
 ethanol oxidation in, 114

Newborns - continued
 regulation of energy intake, 4-5
Niacin, requirements of, 141, 154
Nickel, 131
Nitroblue-Tetrazolium Test, 275
Nitrogen
 balance studies, 16, 23-24, 30, 151, 162, 169-176, 178-188, 202-203, 225, 245, 262
 content in human milk, 167
 intake, 28, 32-33
 losses, 64, 82-83, 166
 requirements for, 209-211, 231, 259, 264
 sources, 259-261
 sparing effect of carbohydrates, 85
Norepinephrine, influence on ketogenesis, 60
Nucleic Acid, metabolism, need for zinc in, 133
Nucleotide synthesis, 13-14
Nutritionist, 159
Oestriol excretion, in placental insufficiency, 211
Oleate, inhibition of gluconeogenesis, 66-67
Opiates, 125
Ornithine, serum levels
 after infusion programs, 174, 190
 in healthy newborns, 207, 210
 in hyperammonemic infants, 262
Osmolality of infusates, 155
Oxidative phosphorylation, 91-92
Oxoacid, role in ketogenesis in liver, 59-60
Oxygen consumption
 after starvation, 149
 during halothane anesthesia, 66
Oxytocin
 release, ethanol effect on, 112
 sensitivity test, 211
Paedamin ^{R}, 173
Palmityl CoA
 oxidation of, 92-94
 synthetase, 89, 92, 93
 transferase, 91
Palmityl Carnitine, 92-95

Pancreatitis, parenteral nutrition in, 215
Pantothenic Acid
  concentration in infusates, 153, 154
  recommendations for, 141
Peptidases, 12
Peptides, digestion of, 12
pH
  blood, during hyperalimentation, 180, 203
  of parenteral solutions, 226, 261
  optimum for ethanol oxidation, 114
Phenylalanine
  estimated requirements, 28, 31, 183
  serum levels after alimentation, 174-175, 188-190
  serum levels in healthy newborns, 31-32, 207-210
Phosphatase, alkaline, abnormalities in, 215-216
Phosphate
  in alimentation solution, 152, 156, 167, 180, 181, 200, 205, 216
  monitoring of, 241, 265-266
  role in sorbitol metabolism, 71-73
Phospho-Enolpyruvate Carboxykinase, 63
Phosphofructoaldolase, hepatic, in sorbitol metabolism, 71-72
Phosphofructokinase, 71-72
Phosphorus
  absorption, 132, 153
  accumulation of, in skeletons of newborns, 166-167
  metabolism, disorders in, 257-260, 265-266
  plasma values after alimentation, 154, 162, 169-171, 204
Phytate, 132
Pigment, changes from fat infusion, 248
Placenta
  amino acid level in, 264

Placenta - continued
  ethanol transfer through, 113
  insufficiency, 206-213
  relation to infant size, 4
Plasma
  as free amino acid pool, 23
  osmolality, 40, 41, 43-44
  post-operative administration of, 162
  replacements, in burned children, 248
Pleural injury, from central venous catheter, 251
Polyols, 152, 181
Polysomes, role in synthesizing proteins, 13-14
Polyvinyl catheter, 155-157, 160, 162, 250
Portal Venous System, role in fat absorption, 89
Positive End Expiratory Pressure, cause of water overload, 236-237
Postoperative
  fluid administration, 55, 165-177, 178
  glucose intolerance, 85, 258
  serum insulin levels, 65
Potassium
  changes during alimentation, 162, 169-173, 204
  in parenteral solution, 149, 152, 156, 167, 179, 180-181, 240
  metabolism, disorders in, 257-259, 265-266
Pregnancy, food intake during, 6, 9
Premature infants
  amino acid requirements, 206
  metabolic complications, 256-268
  parenteral nutrition of, 163, 178-198, 233, 235
Progesterone, influence on protein intake, 7
Proline
  concentrations in infusates, 182
  requirements, 192

Proline - continued
   serum levels after infusion, 174
   serum levels in newborns, 207-210

Protein
   caloric value of, 88-89
   endogenous source, 12-13
   fetal deposition of, 206
   intake, 6, 143, 263
   malnutrition, 49
   requirements for, 27-36, 182
   role in alcohol dehydrogenase activity, 115
   role in amino acid utilization, 11-26
   role in hepatic gluconeogenesis, 71
   role in intravenous nutrition, 135, 146-149, 151, 152, 180, 237
   serum values after infusion, 154
   sparing by fat infusions, 182
   synthesis, 56, 131, 133, 149, 151-152, 172, 185, 194, 225

Pseudomanas, sepsis, 244 (also see bacteria)
Pulmonary insufficiency, 199, 203
Pump, for infusion, 157, 158, 241
Purines, 13-14
Pyrexia, 147
Pyridoxine
   requirements, 141, 154
   urinary excretion of, 140
Pyruvate
   -lactate ratio in livers, 114
   levels after carbohydrate infusion, 77-79
   role in alanine re-synthesization, 65
Radiologic studies, see X-rays
Rash, in lipid deficiency, 264
Rehabilitation, of undernourished children, 146-150
Renal
   capacity of, 119-120, 128
   disease, 85, 225-230, 232-233, 235

Renal - continued
   failure intravenous diet, 225-229
   insufficiency, 121
   preference for fatty acids, 88
   tubules, regeneration of, 227
Reproductive development, 132
Respiratory Distress Syndrome
   dextrose infusion in, 235-236
   fluid tolerance in, 233
   glucose tolerance in, 47, 49, 199
Respiratory quotient, 245
Riboflavin
   in cell replication, 131
   in infusates, 139-140
   requirements for, 141, 154
   (also see vitamins)
Ribonucleic Acid (RNA), 13-14
Rickets, 140, 153
Roentgenography (see X-ray)
Round cell infiltrate, 216-219
Rubella Syndrome, 233
Saline, intravenous use, 124, 161, 181, 232, 260
Selenium, 131
Seizures, 125
Sepsis
   alimentation during, 235
   associated with IV lines, 159, 204, 231, 242-246, 256, 269-278
   burn wound, 239
   glucose tolerance during, 47
Serine
   conversion of glycine to, 194
   serum levels after alimentation, 174
   serum levels in newborns, 207, 210
Serum
   osmolality, 154-155
   water concentration of, 122
SGOT (see GOT)
SGPT (see GPT)
Shock, central venous lines in, 248, 262
Short Gut Syndrome, 162
Silicon, requirements, 131
Silicone, catheters, 155-157

SUBJECT INDEX

Skin
  irritation at IV site, 255
  role in ethanol elimination, 113
Sodium
  balance, 162, 256-259
  concentration in infusates, 149, 152, 156, 167, 180, 181, 240
  influence on water regulation, 124, 127
  metabolism, disorders of, 257-259, 266
  overload, 166
  plasma levels after alimentation, 204, 265
Solute, output and urine flow, 123-124
Sorbitol
  use in infusates, 152, 247
  utilization of, 71-86, 181
Spinal fluid, ethanol in, 113
Staphylococcus sepsis, 244
Starvation
  effect of halothane during, 67
  ketone body utilization in, 59-60
Steatosis, 218
Sterile precautions, 238
Steroids, influence on NBT test, 275
Streptozotocin diabetics, 77
Stool
  normalization of, after alimentation, 220-223
  water losses in, 128
Stress
  calorie requirements during, 239
  glucose intolerance during, 258
Succinate, role in ketogenesis, 59
Succinyl-CoA, 59
Sucrose, effect on uric acid metabolism, 80
Surgery
  fluid tolerance following, 232-233

Surgery - continued
  parenteral nutrition in, 98-110, 125-126, 214-220, 223, 233, 235, 265
Taste, in zinc deficiency, 133
Taurine, serum levels, 174
Team approach, 159, 265, 267
Tetany, in hypocalcemia, 260
Thiamine (also see vitamins)
  half life, 139
  recommendations for, 141, 154
  urinary excretion of, 140
Threonine
  in infusates, 183, 190, 226
  requirements of, 28, 31
  role in SGPT activity, 216
  serum levels after infusion, 174
  serum levels in newborns, 207-210
Threonineserine Dehydratase, 14
Thrombocytopenia, following use of Intralipid^R, 237
Thrombophlebitis, 231, 248
Thrombus, formation, 155, 231, 239, 242-246, 248, 253, 255
Thymidine, 133
Tin, required trace element, 131
Transaminases, serum, 215, 257-262, 265 (also see GOT, GPT)
Transketolase
  effect of vitamins on, 143
  role in xylitol metabolism, 72
Trace elements, 156
Trauma
  catabolism following, 65
  glucose and fat tolerance after, 65-66
Tricarboxylic Acid Cycle, 59
Triglycerides
  hydrolysis of, 89
  medium chain, 237
Triokinase, in phosphorylation of glyceraldehyde, 72
Tryptophan
  in intrauterine feeding, 209
  in parenteral solution, 183, 226
  requirements, 28, 31
  tolerance and fate of, 15-18

Tyrosine
  in parenteral fluids, 183
  liver uptake of, 13
  oxydase system, 190
  serum levels after infusion, 174-175
  serum levels in newborns, 207-210
Tyrosinemia, 214
Undernutrition, 7, 146
Urea
  excretion, 83, 124, 125, 232, 262-263, 266
  in amniotic fluid, 209
  serum levels after infusion, 184
  synthesis, 13, 19, 65, 66
Uremia
  positive nitrogen balance in, 172, 225
  utilization of fructose in, 55, 250
Uric Acid
  after carbohydrate infusion, 77, 79, 80, 85
  concentration in amniotic fluid, 209
  synthesis, 56, 71
Urine
  changes after prolonged nutrition, 136
  ethanol in, 113
  excretion of estriol in, 211
  excretion of nitrogen, 169
  hydroxyproline concentrations in, 192
  maintenance of volume, 119, 120, 122, 231-233, 241, 256
  3-methylhistidine in, 20
  phosphate loss through, 259
  studies during alimentation, 151, 154-155, 159, 178-190, 232, 241, 265-266
  vitamins in, 141, 143
Uterus, protein requirement for growth, 6
Uxlulose, role in oxidation of xylitol, 72

Valine
  in Maple Sugar Urine Disease, 237
  in parenteral solution, 183, 190, 226
  requirements, 28, 31
  serum values, 15, 174, 188, 207-210
Vanadium, as required trace element, 131
Vasopressin
  inappropriate release of, 125, 236
  role in urine concentration, 124, 127
Veins
  central, 155-163, 250-255
  central venous pressure, 155
  periferal use, 55
Venticular Septal Defect, 228-229
Vim Silverman Needle, for directing catheters, 156-157
Vitamins
  in parenteral fluids, 131-145, 151-156, 162, 181, 200, 216, 240
  metabolism, disorders in, 257, 260, 264
  requirements, 154, 168, 173
  toxicity, 140
Water
  balance, 154, 256-259
  post partum mobilization of, 161
  requirements and tolerance, 119-129, 180, 231, 232, 234-238
Weaning, from parenteral nutrition, 259
Weight
  gain, 4-5, 30, 120, 147-148, 161, 172, 185-187, 192, 220-223
  loss, 241
  low birth weight, 39-53, 199-205
  role of hypothalamus, 7
  routine determinations of, 154, 180

SUBJECT INDEX

Whirlpool debridement, 239
World Health Organization, 27, 29, 32
Wound healing
  in zinc deficiency, 132
  of burns, 239-246
  of surgical incisions, 161
X-Rays
  for catheter placement, 156-157, 241, 246, 254
  gastrointestinal, 161, 220-223, 242
Xylitol
  dehydrogenase, 72
  in parenteral solutions, 152
  tolerance in newborns, 181
  utilization of, 71-86
Yeast, sepsis, 242-246, 270-275
Zeiss Loop, for broken catheter removal, 252
Zinc
  deficiencies, 264
  in parenteral solution, 152
  role in protein synthesis, 131-138